Study (

for Joseph F. Healey's

Statistics

A Tool for Social Research

Fifth Edition

Prepared by
Rebecca Davis
College of Notre Dame of Maryland

Wadsworth Publishing Company
I(T)P® An International Thomson Publishing Company

Belmont, CA • Albany, NY • Boston • Cincinnati • Johannesburg • London • Madrid • Melbourne
Mexico City • New York • Pacific Grove, CA • Scottsdale, AZ • Singapore • Tokyo • Toronto

Printed in Canada.
1 2 3 4 5 6 7 8 9 10

For more information, contact Wadsworth Publishing Company, 10 Davis Drive, Belmont, CA 94002, or electronically at http://www.wadsworth.com.

International Thomson Publishing Europe
Berkshire House
168-173 High Holborn
London, WC1V 7AA, United Kingdom

Nelson ITP, Australia
102 Dodds Street
South Melbourne
Victoria 3205 Australia

Nelson Canada
1120 Birchmount Road
Scarborough, Ontario
Canada M1K 5G4

International Thomson Publishing Japan
Hirakawa-cho Kyowa Building, 3F
2-2-1 Hirakawa-cho
Chiyoda-ku
Tokyo 102, Japan

International Thomson Editores
Seneca, 53
Colonia Polanco
11560 México D.F. México

International Thomson Publishing Asia
60 Albert Street #15-01
Albert Complex
Singapore 189969

International Thomson Publishing Southern Africa
Building 18, Constantia Square
138 Sixteenth Road, P.O. Box 2459
Halfway House, 1685 South Africa

ISBN 0-534-55262-5

CONTENTS

PREFACE

This text is designed as a supplement to Healey, Statistics: A Tool for Social Research, Fifth Edition, 1999. It is not designed as a replacement of your text. Instead, this text provides multiple-choice questions, additional work problems, and detailed answers for each chapter. I strongly suggest using this book only as a supplement to the text.

This study guide was written to maximize your understanding of the concepts in your main textbook. After you have finished studying for an exam, try the multiple-choice problems at the end of each chapter. The answers are at the back of the book, so you can check your answers. These questions will measure your understanding of the general concepts of each chapter.

Next, try your hand at the work problems. These problems will give you additional practice with data and word problems. Detailed answers to all of the problems are at the end of each chapter. Remember, the more practice you get, the easier it will be!

Many of the examples in this text come from contemporary social problems, but all of the data are fictional. This text is written to be easy to read and easy to understand. Statistics has a bad reputation, but if you sit back and relax a bit, you might find you even enjoy it!

My ability to write this text comes from the many hours I have spent explaining the fundamentals of statistics to my students at the University of Maryland, College Park, Bowie State University, and most recently at the College of Notre Dame of Maryland.

A special thanks to Tom Valente who has helped me become a better teacher.

CHAPTER ONE
INTRODUCTION

I recently took an introductory French course at the university and I was struck by how similar my learning experience was to that of many of my students. The anxiety I felt before each class, the endless memorization of new terms, and the constant learning of new concepts that I could not get my mind around, are all things I have heard first year statistics students talk about. It occurred to me that learning statistics is like learning a new language and should be approached as such. I realized that there is a structure and a vocabulary with regional differences that must be learned, but most importantly like any new language, you cannot be expected to be fluent in it after one semester of introduction.

Likewise, after you finish this course, you will not be expected to be fluent in statistics. Instead, you will likely have a better understanding of the structure of statistics, which tests to use with which variables, and which tests are appropriate to use when.

Many of you will struggle with the course, the same way I struggled with my introductory French course. But in the end you will have a better understanding of research methodologies and statistical tests, and be ready to venture on to a more advance level.

It is important to remember that statistics is a language all its own, with its own properties and slang. Remind yourself of this fact often, especially when you are having trouble with a phrase or word ordering: you are learning a new language.

Also, keep in mind that statistics is <u>not</u> mathematics. We use mathematical operations to compute the statistics, but the statistics themselves are tools for describing and defining populations of <u>people</u>. I emphasize the word <u>people</u> because that is what we are most often studying: people, groups, societies; such as what is the average height of basketball players in the NBA? What is the average salary for college graduates?

There are two types of statistics which make up the majority of your text book. The first, **descriptive statistics**, are simply measures for describing groups on some variable. The second type of statistic is **inferential statistics** that allow us to <u>infer</u> our results from a sample to the rest of the population.

A few new words have already appeared:

<u>Variable</u>: any trait that can change value from case to case (Healey, 1999: 3), such as race: I may be white, my neighbor may be black. The variable "race" can change from one case (mine) to the next (my neighbor).

<u>Population</u>: the total collection of all cases in which the researcher is interested (Healey, 1999: 9). Perhaps you want to study females of childbearing age in the U.S. Your population then is <u>every</u> female between the ages of 13 and 45 in the United States.

<u>Sample</u>: a carefully chosen subset of a defined population (Healey, 1999: 9). It is virtually impossible to ask every single person in a population his or her sex, weight, height, or whatever else you are interested in studying, therefore you need a <u>sample</u>, or a subset, of the population.

Variables, then, are the items on a questionnaire or survey which we are using to measure our concepts, and are categorized into three types, or "levels" of measurement: nominal, ordinal, and interval-ratio[1].

If I could pick only one concept that I feel to be most important, it is levels of measurement. If you can look at a variable and know it's level of measurement, then you can figure out what test to use on it. But, as you will see later, there are different tests for different levels of measurement. Picking the wrong test not only will give you incomplete information, but in many cases it will give you inaccurate information.

Let's review the three levels of measurement:

Nominal: Simply classifying cases into categories, such as male or female. There is no ranking or ordering of the categories; you cannot say that males are higher than females, or that Protestants are higher than non-Protestants, only that a person can be classified or put into one and only one category.

The categories for nominal variables must be *mutually exclusive* (each subject fits into one and only one category), *exhaustive* (there must be a category which every subject can fall in to), and *homogeneous* (the categories must be similar enough to allow generalization). Healy (1999: 11) gives the example of scales for religious categories. In the first scale, a person who is Episcopalian could fall into more than one category: Episcopalian and Protestant. Therefore, that categorization is not mutually exclusive. In the second scale, a person who is Muslim would not fall into any of the categories, therefore the scale is not exhaustive. In the third scale, those people who would fall into the category "Non-Protestant" (which would include Catholics, Muslims, Buddhists, Nones) are so unsimilar in respect to religious beliefs, that this categorization is not homogeneous. Examples of nominal variables are: gender, religious preference, all Yes/No questions.

Ordinal: Ordinal variables have all the same qualities of nominal variables, but also go one step beyond -- the categories are ranked or ordered from high to low, such as class standing: are you a freshman, sophomore, junior, or senior? Not only can you be classified into one and only one category, but we can rank-order these categories based on seniors having more education and/or unit hours than juniors, etc. Note that there is no true zero point, meaning that we cannot say that being a senior is three times higher than being a freshman, only that being a senior is higher than being a freshman.

Examples of ordinal variables are: class ranking, social class, most scales, especially those ranked from "strongly agree" to "strongly disagree".

Interval-Ratio: Interval-ratio variables go even a step beyond ordinal variables. These variables are classified into categories (as are nominal variables), and are rank-ordered (as are ordinal variables), but most importantly, the exact distance between the categories is known. For example, if you were to classify all of your classmates as either freshmen, sophomores, juniors, or seniors, we would have an ordinal measure. But if you were to classify them by the exact number of credits or units they have completed, we would have an interval-ratio variable. Thus you could determine that someone who has completed 64 unit hours has twice as many unit hours as someone who has completed only 32 hours. Therefore, there are actual real numbers associated with each category.

---

[1] As noted in Healey (1999: 13), many statisticians distinguish between interval level and ratio levels of measurement. The difference is that interval level does not have a true zero point, whereas ratio level has a true zero point. At this point in your learning, though, it is easier to consider these two levels as one: interval-ratio.

Examples of interval-ratio variables are: age, income, years married, number of children, years of education.

It is important to note that these levels of measurement are based on a hierarchy: nominal-level variables yield the least amount of information; they simply place each person into a category. Ordinal-level variables yield more information because each person or case is placed into a category and each category can be ranked from high to low. Interval-ratio-level variables yield the most information: not only are these categories ranked, but the exact distance between the ranks is known and can be measured, thus permitting mathematical operations to be performed.

MULTIPLE-CHOICE QUESTIONS

1. All of the boys in junior high school is an example of a _________.
   a. sample
   b. population
   c. variable
   d. level of measurement

2. Two types of statistics are______________________.
   a. informational and deceptive
   b. descriptive and informational
   c. inferential and deductive
   d. descriptive and inferential

3. Which of the following is an example of a sample?
   a. A college dorm
   b. The President of the university
   c. A subset of the university
   d. All students in all colleges across the country

4. A researcher is interested in drinking behavior at a university. What is the population?
   a. The university
   b. The administration
   c. The carefully selected subset she uses for her study
   d. The questions she asks regarding drinking behavior

5. Using the same example from Question 4 above, what variable might she use?
a. The student body
b. A subset of the student body
c. Amount of alcohol one drank during the past month
d. Descriptive statistics

6. The variable "social class" is a(n) ____________ level variable.
a. nominal
b. ordinal
c. interval-ratio
d. none of the above

7. The variable "Do you favor or oppose abortion" is a(n) ____________ level variable.
a. nominal
b. ordinal
c. interval-ratio
d. none of the above

8. If you were to rank the levels of measurement from lowest to highest by the amount of information each one yields, what order would you put them in?
a. Ordinal, interval-ratio, nominal
b. Interval-ratio, ordinal, nominal
c. Nominal, ordinal, interval-ratio
d. Nominal, interval-ratio, ordinal

9. An adequate scale for measuring a nominal-level variable should be ________________.
a. mutually exclusive, exhaustive, and heterogeneous
b. mutually exclusive, exhaustive, and homogeneous
c. mutually inclusive, exhaustive, and homogeneous
d. none of the above

10. The variable "Number of children" is a(n) ____________ level variable.
a. nominal
b. ordinal
c. interval-ratio
d. none of the above

11. Which of the following scales would be best for the variable "ethnicity"?
a. White, black, other
b. White, black, hispanic
c. Black, non-black
d. White, black, hispanic, asian, other

12. Descriptive statistics are used to ________________________.
a. describe a population
b. describe the relationship between statistics and social science
c. describe a distribution of a variable
d. describe the textbook

13. Inferential statistics are used to ________________________.
a. infer the population results to the sample
b. infer the sample results to the population
c. infer our knowledge of statistics to social science
d. infer our knowledge to the instructor

14. A discrete variable is ______________________________.
a. a variable that is identified as a causal variable
b. a variable with a unit of measurement that can be subdivided
c. a variable with a basic unit of measurement that cannot be subdivided
d. a variable that is identified as an effect variable

15. A continuous variable is ______________________________.
a. a variable that is identified as a causal variable
b. a variable with a unit of measurement that can be subdivided
c. a variable with a basic unit of measurement that cannot be subdivided
d. a variable that is identified as an effect variable

WORK PROBLEMS

Below are a few of the variables from the 1996 General Social Survey data set (Appendix G in your textbook). What is the level of measurement for each variable?

1. Last week, were you working full-time, part-time, going to school, keeping house, or what?
1. Working full-time
2. Working part-time
3. With a job but not at work (ill, vacation, etc)
4. Unemployed
5. Retired
6. In school
7. Keeping house

Level of Measurement ____

2. Are you currently married, widowed, divorced, separated, or have you never been married?

1. Married	4. Separated
2. Widowed	5. Never married
3. Divorced	6. NA

Level of Measurement ______________________________

3. How many children have you ever had? Please count all that were born alive at any time (including any from a previous marriage).
   0 - 7. Actual number
   8. Eight or more
   9. NA

Level of Measurement ______________________________

4. Respondent's highest degree
   0. Less than HS
   1. High school
   2. Assoc/Jr. college
   3. Bachelor's
   4. Graduate
   9. NA or DK

Level of Measurement ______________________________

5. Race of respondent
   1. White
   2. Black
   3. Other

Level of Measurement ______________________________

6. Highest year of school completed
   0-20 Actual number of years
   98. Don't know or NA

Level of Measurement ______________________________

7. Respondent's gender
   1. Male
   2. Female

Level of Measurement ______________________________

8. Do you think the use of marijuana should be legal or not?
   1. Should
   2. Should not
   0. Not Applicable, DK, NA

Level of Measurement ______________________________

9. Do you strongly agree, agree, disagree, or strongly disagree that methods of birth control should be available to teenagers between the ages of 14 and 16 if their parents do not approve?
1. Strongly agree
2. Agree
3. Disagree
4. Strongly disagree
0. Not Applicable, DK, NA

Level of Measurement ________________________________

10. On the whole, how satisfied are you with the work you do -- would you say that you are very satisfied, moderately satisfied, a little dissatisfied, or very dissatisfied?
1. Very satisfied
2. Moderately satisfied
3. A little dissatisfied
4. Very dissatisfied
0. Not applicable, DK, NA

Level of Measurement ________________________________

DETAILED ANSWERS

1. This variable is ordinal because it can be ordered by the amount of time the respondent was employed within the prior week. We know that working full-time consists of more employed hours than working part-time, etc.

This variable is a bit sticky, though, because being retired (category 5) is not more employment than being in school (category 6), which is not necessarily more employment than keeping house (category 7). Thus, some researchers might choose to consider this variable as nominal. This is fine, but as you will see later, we can gain more information and insight from the higher level of measurement, therefore it would be advantageous to us to keep this variable as ordinal.

One solution then, would be to use only those categories which we can accurately place in a hierarchy: working full-time, working part-time, unemployed.

2. This variable is nominal because these categories cannot be ordered or ranked. Being married is not higher than not being married. Therefore, each respondent can be placed in one and only one category, of which there is no value placed.

3. Because the question asks for the actual number, this variable is interval-ratio. We know that having two children is two times more than having one child; the distance between the numbers is known. As such, this variable yields the most information and mathematical operations can be performed on this type of variable.

4. This variable is ordinal, again because we can rank-order the categories. We know that having a college degree is higher than having a high school diploma, but we cannot say for certain (with the amount of detail given in these categories) that having a bachelors degree is four times higher than having a high school diploma.

If we want to get this type of information, we can transform these categories into their corresponding years of education, such as high school diploma entails 12 years of education, associate degree entails 14 years of education, etc. Unfortunately, though, in doing so you will have to assign values for those unknowns such as "less than high school" and "graduate degree". Someone with less than high school might have five years of education, while others might have eleven. Be careful of these problems whenever you attempt to change the level of measurement--you might lose more data than it's worth!

5. Race of respondent is nominal level, because these are simply categories: they cannot be ranked or ordered, but only placed into categories.

This variable could be improved, though, prior to administering the questionnaire by adding in other categories. Think of all the different racial and ethnic groups in the U.S. It is hard to imagine them being placed into just three groups. It would be a better variable had the categories been "white", "black", "Hispanic", "Asian", and "other". Since there are many cultural differences between Hispanics and Asians, lumping them together might lose some vital information.

6. This variable asks the respondent the actual number of years, therefore it is measured at the interval-ratio level. Compare this question with Question 4. Can you see that we obtain much more information when we ask the interval-ratio version of this question rather than the ordinal level? Here, we don't have to make "guesstimates" about the number of years, but instead know the value exactly.

7. Gender is a nominal-level variable: being either male or female is simply a categorization, not placing a value on these categories.

8. The legalization of marijuana variable is nominal, because again you are simply placing respondents into one of two categories: do they support legalization or not. We cannot order these categories nor place a value on them.

9. Because this variable asks the respondent to rank how strongly they feel about birth control for teenagers, this variable is ordinal. We can rank-order the categories so those who strongly favor are ranked higher than those who strongly oppose. We are ranking their attitudes: how strongly do they feel about this subject?

10. Satisfaction with job is also ordinal because you are asking the respondents to rank-order their satisfaction.

CHAPTER TWO
BASIC DESCRIPTIVE STATISTICS
Percentages, Ratios and Rates, Tables, Charts, and Graphs

You have now been armed with the basic knowledge necessary to begin data analysis. Keep in mind that data analysis belongs to one of two groups: descriptive (describing the sample distribution of a variable) or inferential (inferring your sample results onto the population). You always want to begin your data analysis by **describing** the distribution of a variable, such as what percentage of your sample is male. Thus we use basic descriptive statistics, such as percentages, ratios, rates, tables, charts, and graphs.

These basic descriptive techniques enable us to display our data graphically in easily readable format, so we can take a quick peek at how each variable looks. The first step in this process is by making a **frequency distribution table**: a table that displays the frequency (or number) of respondents in each category for that variable. We do this with all three types of variables (nominal, ordinal, interval-ratio), although the process is a bit different for interval-ratio variables.

Basically it works like this: let's say you have 100 respondents, 60 of whom are female, 40 of whom are male. To make a frequency distribution table, we want to display this distribution across these two categories, such as the table below. Looking at this table we can see that the majority of the respondents are female (60 out of 100, or 60%). This table, like all frequency distributions, gives not only the frequency, but the percentages and/or the proportion in each category as well.

TABLE 2.1 FREQUENCY DISTRIBUTION OF GENDER

GENDER	FREQUENCY (f)	PERCENTAGE (%)	PROPORTION (p)
(1) Male	40	40%	0.40
(2) Female	60	60%	0.60
N=	100	100%	1.00

To find the percentage for each category, we use:

$$Percentage(\%) = (\frac{f}{N})\,x100$$

where f = frequency, or the number of cases in any category
N = number of cases in all categories

$$Percentage(\%) = (\frac{60}{100})x100 = 60\%$$

From the example above, the percentage of females is:
To find the proportion, we use:

$$Proportion(p) = (\frac{f}{N})$$

where f = frequency, or the number of cases in any category
N = number of cases in all categories

From the example above, the proportion of females is:

$$Proportion(p) = (\frac{60}{100}) = 0.60$$

The only difference between percentages and proportions is their base number: percentages are based on 100 while proportions are based on 1. They yield basically the same information.

Frequency distribution tables are created differently for interval-ratio variables. Think of all the different ages of the other students in your statistics class. If you have a class of 50 students, you probably have some students who are 18 years old, perhaps one or two who are under 18, a few who are 19, several who are 20, some who are between 21 and 24, perhaps some who are around 30, and some who are older than 30. To put all of these scores into a frequency distribution table, you might have quite a few categories. It's possible, then, to break these categories down into more manageable sizes, or **class intervals** such as 17-19 years of age, 20-22, 23-25, 26-28, etc. This would enable you to see the distribution more clearly. The size of the class intervals is up to you, depending on how much information you want to convey.

Remember also that interval-ratio variables are continuous, meaning that all the values in the distribution are connected. In order to maintain the continuity of the variable, you will want to use **real class limits**, which are created by dividing the distance between your stated limits (in the example above 17-19 = two years) in half (1) and adding this value to the upper and lower limit of each category. Therefore, the real class limits for the category 17-19 would be 16-20; for 20 - 22, it would be 19 -23, etc. This step is essential if you plan to graphically display your data in a histogram.

Midpoints are the points exactly in the middle of the intervals, and are found by dividing the sum of the upper and lower real class limits by two. If the real class limits are 16 - 20, the sum is 36; divided by two, the midpoint is 18. Midpoints are used primarily for the purpose of creating frequency polygons, which are described in more detail below.

TABLE 2.2 FREQUENCY DISTRIBUTION OF AGE IN CLASS , WITH REAL CLASS LIMITS AND MIDPOINTS INCLUDED (N=50)

AGES	FREQUENCY (*f*)	CUMULATIVE FREQUENCY	REAL CLASS LIMITS	MID-POINT
17-19	8	8	16-20	18
20-22	21	29	19-23	21
23-25	13	42	22-26	24
26-28	6	48	25-29	25
29-31	2	50	28-32	30
N=	50			

Cumulative frequencies and the **cumulative percentages** are also used in frequency distributions for interval-ratio variables. Cumulative frequencies are the summation of the frequencies down the columns. Thus if there are 8 students between the ages of 17 and 19, the first cumulative frequency would be 8. If there are 21 students between the ages of 20 and 22, then the cumulative frequency would be 8 +21, or 29.

Cumulative percentages work the same way, but you sum the percentage column with each additional category instead of the frequencies.

Both of these measures make it easier to view the distribution. With a quick look at the cumulative percentage column, you can tell where 90% of the cases fall below; with the cumulative frequency column you can quickly find the median for the distribution.

After you display your data in the form of frequency distribution tables, a second way to display your data is with charts and graphs. Like many other measures, the chart or graph you use will depend largely on the level of measurement of your variable. For nominal and ordinal level variables, you may use either **pie charts** or **bar graphs**. For interval-ratio variables, you should use either **histograms** or **frequency polygons**. All of these charts and graphs share the same purpose: to make the distribution of the variable among the categories clearer and easier to read.

Pie charts are the easiest to construct. Make a circle, and divide the circle into slices, as you would on a pie, which correspond to the frequency (or percent) of respondents in each category.

Bar graphs use vertical bars to represent the frequency of cases in each category. Each category is labelled along the horizontal axis on the bottom, and either the frequencies or percentages are marked along the vertical axis on the left-hand side. For each category, make a vertical bar that reaches to the frequency of that category. Remember that for nominal and ordinal variables, the categories are discrete, therefore the bars of each category should not touch each other.

For histograms, you use the real class limits and set the chart up the same way as you would a bar chart, with the frequencies along the vertical axis and the real class limits along the horizontal axis. Because interval-ratio variables are continuous, you want your frequency bars to touch each other, starting and ending at each real class limit.

Similarly, frequency polygons graphically display the distribution of continuous data, but with a line that connects each frequency of the categories at each midpoint. Like bar charts and histograms, frequency polygons display frequencies along the vertical axis, but midpoints of the real class limits along the horizontal axis. Each frequency of the distribution is marked with a dot, then a line is used to connect the dots. This then graphically illustrates the distribution of scores by categories.

All of these methods are used essentially to describe the distribution of scores on some variable. It is important to remember to use the appropriate measure for the appropriate level of measurement.

MULTIPLE-CHOICE QUESTIONS

1. A frequency distribution table is:
   a. a table with the total number of respondents
   b. a table with the number of respondents by category
   c. the percentage or proportion of respondents by category
   d. all of the above

2. For nominal variables, the frequency distribution should have:
   a. cumulative frequencies
   b. cumulative percentages or proportions
   c. percentages or proportions
   d. all of the above

3. For ordinal variables, the frequency distribution should have:
   a. cumulative frequencies
   b. cumulative percentages or proportions
   c. percentages or proportions
   d. all of the above

4. For interval-ratio variables, the frequency distribution should have:
   a. cumulative frequencies
   b. cumulative percentages or proportions
   c. percentages or proportions
   d. all of the above

5. Percentages are based on:
   a. one
   b. five
   c. fifty
   d. one hundred

6. Proportions are based on:
   a. one
   b. five
   c. fifty
   d. one hundred

7. Ratios are:
   a. the number of actual occurrences of some phenomenon or trait divided by the number of possible occurrences per some unit of time
   b. the number of cases in one category divided by the number of cases in another category
   c. the number of cases in one category of a variable divided by the number of cases in all categories of the variable
   d. The horizontal dimension of a table

8. Rates are:
   a. the number of actual occurrences of a phenomenon or trait divided by the number of possible occurrences per a unit of time
   b. the number of cases in one category divided by the number of cases in another category
   c. the number of cases in one category of a variable divided by the number of cases in all categories of the variable
   d. the horizontal dimension of a table

9. To graphically display the distribution of a nominal variable, you should use a:
   a. pie chart or bar graph
   b. pie chart or histogram
   c. bar graph or histogram
   d. histogram or frequency polygon

10. To graphically display the distribution of an ordinal variable, you should use a:
   a. pie chart or bar graph
   b. pie chart or histogram
   c. bar graph or histogram
   d. histogram or frequency polygon

11. To graphically display the distribution of an interval-ratio variable, you should use a:
   a. pie chart or bar graph
   b. pie chart or histogram
   c. bar graph or histogram
   d. histogram or frequency polygon

12. A histogram uses __________ data.
   a. dichotomous
   b. continuous
   c. discrete
   d. contaminated

13. A bar chart uses __________ data.
   a. dichotomous
   b. continuous
   c. discrete
   d. contaminated

14. In a frequency polygon, frequencies are graphed with the use of _____________.
   a. vertical bars
   b. horizontal bars
   c. real class limits
   d. midpoints

15. In a histogram, frequencies are graphed with the use of _____________.
   a. vertical bars
   b. horizontal bars
   c. real class limits
   d. midpoints

WORK PROBLEMS

1. A researcher collected data on attitudes toward abortion. Her first step in analyzing the data was to make the frequency distribution table below.

Attitude Toward Abortion	f
Strongly favor	42
Favor	57
Oppose	52
Strongly oppose	43
	194

a. Complete the table.
b. Graphically display the distribution.
c. What conclusions can you make regarding her data?

2. The data below represent the age at which a sample of newlyweds got married.

17	25	27	22
18	18	24	31
24	19	22	27
22	20	21	23
36	19	19	22
35	23	27	28
16	18	26	21
21	23	25	22

a. Make a frequency distribution table from this data.
b. Graphically display the data.
c. What conclusions can you make?

3. The data for age first married was broken down by gender.

Males		Females	
24	27	17	18
36	23	18	23
21	19	22	19
25	26	21	27
19	25	35	22
23	22	16	21
27	28	18	22
24	31	19	20

a. Make a frequency distribution table for each group (male and female) from this data.
b. What conclusions can you make?

4. The following questions were asked of 20 respondents.

Sex: 1 Male 2 Female

Education (given in actual years)

Should marijuana be legalized?
1 Yes
2 No

The data for these question are below:

Case	Sex	Education	Legalize Marijuana
1	1	12	1
2	2	12	2
3	1	14	2
4	1	15	1
5	2	11	1
6	2	9	2
7	1	16	1
8	1	16	1
9	2	14	2
10	2	12	1
11	2	11	1
12	1	12	1
13	1	12	2
14	1	17	1
15	2	18	2
16	1	10	2
17	1	14	1
18	2	15	2
19	1	9	2
20	1	11	1

a. Construct a frequency distribution for each of the variables.
b. Graphically display each of the distributions.

5. The following data report the overall satisfaction with the police response in one inner-city neighborhood.

Satisfaction w/ police	f
Very satisfied	25
Satisfied	32
Dissatisfied	38
Very dissatisfied	27
	122

a. Complete the frequency distribution table.
b. What conclusions can you make regarding satisfaction with police?

6. The following data are from a quality of life scale administered to hospitalized respondents. Low scores indicate low quality of life, high scores indicate high quality of life. Assuming the data to be interval-ratio, construct a frequency distribution and write a conclusion.

Quality of Life

29	22	25	21	15
25	15	19	29	29
30	29	10	9	22
17	13	27	12	7
19	20	9	30	19
28	17	24	18	24

7. The city council is trying to get an idea of the religious background of the residents. The following data were collected. What would you conclude about their findings?

Religious Affiliation	f
Protestant	240
Catholic	467
Jewish	139
Other	86
None	22
	954

8. Construct a frequency polygon with the following data:

Years of Education	f
7-9	19
10-12	57
13-15	36
16-18	22
19-21	10
	144

9. Graphically display the following data using percentages and a pie chart.

Highest Educational Degree	f
Less than HS	8
High school	27
Assoc/Jr college	15
Bachelor's	12
Graduate	3
	65

10. In a city of 750,000 people, the frequency of unwed pregnancies in a one-year period was 1,875. What is the crude unwed pregnancy rate for this city?

11. The following data about drinking patterns were collected at a large university.

		Major	
Frequency of Drinking	Social Science	Arts and Humanities	Physical Science
Less than once/week	12	8	6
Once or twice/week	17	5	8
Three-five times/week	13	4	6
Five-seven times/week	6	2	3
More than seven/week	2	0	2
	50	19	25

a. Students in which major drink proportionally more than the other majors?

b. What is the ratio of heavy drinkers (seven or more times per week) to light drinkers (less than once per week) by majors?

c. Make a bar chart which includes all three majors by the amount of drinking per week.

12. In a city with a population of one million, there were 516 homicides in the past year. What is the crude homicide rate?

DETAILED ANSWERS

1a.

Attitude Toward Abortion	f	%
Strongly favor	42	22
Favor	57	29
Oppose	52	27
Strongly oppose	43	22
	194	100%

b. Use either a pie chart or bar graph to display either the frequencies or the percentages in each category.

c. The majority of respondents is not strongly for or against, but fall fairly evenly between the two middle categories.

2a.

Age first Married	f	%	cumulative f	cumulative %
16-17	2	6	2	6
18-19	6	19	8	25
20-21	4	12.5	12	37.5
22-23	8	25	20	62.5
24-25	4	12.5	24	75
26-27	4	12.5	28	87.5
28-29	1	3	29	90.5
30 +	3	9	32	99.5
	32	99.5%		

b. To graphically display this data, you need to use either a frequency polygon or a histogram because age is an interval-ratio variable. Thus you need to find the real class limits and midpoints for each category. For the categories above, they would be as follows:

Real Class Limits	Midpoints
15.5-17.5	16.5
17.5-19.5	18.5
19.5-21.5	20.5
21.5-23.5	22.5
23.5-25.5	24.5
25.5-27.5	26.5
27.5-29.5	28.5
29.5-31.5	30.5
31.5-33.5	32.5
33.5-35.5	34.5
35.5-37.5	36.5

c. There are more people who got married between the ages of 22 and 23 than any other category (25%), and the majority of the sample (62.5%) was married by the age of 23.

3.

MALES

Age first Married	f	%	cumulative f	cumulative %
19-20	2	12.5	2	12.5
21-22	2	12.5	4	25
23-24	4	25	8	59
25-26	3	19	11	69
27-28	3	19	14	88
29-30	0	0	14	88
31-32	1	6	15	94
33-34	0	0	15	94
35-36	1	6	16	100
	16	100		

FEMALES

Age first Married	f	%	cumulative f	cumulative %
16-17	2	12.5	2	12.5
18-19	5	31	7	43.5
20-21	3	19	10	62.5
22-23	4	25	14	87.5
24-25	0	0	14	87.5
26-27	1	6	15	93.5
28-29	0	0	15	93.5
30-31	0	0	15	93.5
32-33	0	0	15	93.5
34-35	1	6	16	99.5
	16	99.5		

b. The data in this sample indicate that the majority of males marry at an older age than do the females. Over half of the female respondents (62.7%) were married by the age of 21, compared to only 25% of males at age 22.

4.

Sex	f	%
Male	12	60
Female	8	40
	20	100

Education	f	%	cf	c%
9-10	3	15	3	15
11-12	8	40	11	55
13-14	3	15	14	70
15-16	4	20	18	90
17-18	2	10	20	100
	20	100%		

Legalize Marijuana	f	%
Yes	11	55
No	9	45
	20	100

b. Since education is interval-ratio level, it should be graphically displayed with either a histogram or frequency polygon. The other two variables are nominal level, and should be graphically displayed with a pie chart or bar graph.

5.

Satisfaction w/ police	f	%
Very satisfied	25	21
Satisfied	32	26
Dissatisfied	38	31
Very dissatisfied	27	22
	122	100

b. There is no on consensus on attitudes toward police response, although more respondents are dissatisfied than satisfied.

6.

Quality of Life	f	%	cf	c%
7-9	4	13	4	13
10-12	2	7	6	20
13-15	3	10	9	30

16-18	3	10	12	40
19-21	5	17	17	57
22-24	4	13	21	70
25-27	3	10	24	80
28-30	6	20	30	100
	30	100		

The majority of the respondents scored lower than 21 on the Quality of Life scale (57%), but 20% of the respondents scored between 28 and 30 on the scale.

7. Almost half of the sample is Catholic (49%). Another 25% is Protestant.

8. Using the midpoints for each category, plot the frequency of each. The midpoints should be: 8, 11, 14, 17, and 20.

9.

Highest Educational Degree	f	%
Less than HS	8	12
High school	27	42
Assoc/Jr college	15	23
Bachelor's	12	19
Graduate	3	5
	65	101

Make a wedge or "slice" of pie equivalent to the percentages in each category.

10. In order to find the rate of unwed pregnancies divide the total number of unwed pregnancies by the population, and multiply by 1,000:

Crude pregnancy rate = (1850/750,000) x 1000 = 2.5

Therefore there are 2.5 unwed pregnancies for every 1,000 people.

11a. In order to find which major drinks more, you need to calculate the proportions for each category by each major. Proportionally, physical science majors drink more than seven times per week (0.8) than do social science (0.04) and arts/humanity majors (0).

b. For every light social science drinker, there is 0.16 heavy drinkers (ratio = 2/12); there are no heavy drinkers in the arts/humanity majors; and for the physical sciences, for every light drinker there is 0.33 heavy drinkers (ratio = 6/25).

c. In order to make a bar chart which includes all three majors per category, you need to have three connected bars per category, each one representing one major in that category.

12. Crude homicide rate = (516/1,000,000) x 10,000 = 5.16

Therefore there are 5.16 homicides for every 10,000 people.

CHAPTER THREE
MEASURES OF CENTRAL TENDENCY

Chapter three of your book introduces you to one of the most important and useful measures in descriptive statistics: measures of central tendency.

"Central tendency" simply means the most common, most central, or the average category or value. Basically you are trying to find the average case: Are males more common in your sample than females? What is the predominant religious category? What is the average age of your sample?

Remember: measures of central tendencies are descriptive measures, so they are only able to *describe the sample*, not the population.

Three measures are used to measure central tendency, and like many other measures, which one to use depends largely on the level of measurement of the variable you are interested in describing[1]. For nominal data, the mode is the only appropriate measure; for ordinal data, the mode is appropriate, but the median is the best measure; and for interval-ratio data, both the mode and the median are appropriate, but in most cases the mean gives you the most information.

The <u>mode</u> is simply the most frequently occurring category. Remember: for nominal data, there is no <u>value</u> associated with each category, only the number of respondents in each category. Therefore if there are 56 males and 47 females, then the **mode** is the **category** "males", not the frequency of 56. "Eyeballing" the distribution is the best way to find the mode: simply find the category with the most number of cases. In some distributions, there will not be <u>one</u> mode, but perhaps two or more modes. If this is the case, then report it as such. State that there is not one most frequently occurring category, but several.

The <u>median</u> is simply the exact <u>center</u> of the distribution. Unlike the mode, it is affected by the distribution, but is not sensitive to extreme values in the distribution the way the mean is.

To calculate the median you can simply find the middle-most score: order the scores and find the one in the middle. If you have an odd number of cases, divide N in half and round up to the nearest number. For example, if you have 31 respondents in your sample, the 16th case would be in the exact center of your distribution, with half of the cases falling above and half falling below.

If you have an even number of cases, the middle-most scores will be N/2 **and** N/2 + 1. In Table 3.1 below, the median would be 20/2 **and** 20/2 + 1, or the 10th **and** 11th cases, which would be in the category "Juniors".

---

[1] If you are unsure of the level of measurements, review section 1.6 (pp. 10-15) of your textbook.

TABLE 3.1 CLASS RANKINGS IN SOCIOLOGY STATISTICS COURSE

Class Ranking	**Frequency (f)**	**Cumulative Frequency**	**Percentage (%)**
Freshmen	3	3	15
Sophomores	5	8	25
Juniors	8	16	40
Seniors	4	20	20
Total	20		100 %

If you have interval-ratio data, though, the median would be [N/2 + (N/2 + 1)]/ 2, or the **average** of the **value** of the 10th and 11th case. For example, the data below represent ages of a study group in a statistics class.

$\underline{X_i}$
18
18
19
19<------------- 4th case
20<------------- 5th case
20
20
20
21

The median then is the average of the values associated with the 4th and 5th cases, or (19 + 20)/2 =19.5. Therefore, the middle-most score is 19.5 years.

The mean is the arithmetic average of scores and is calculated by summing all of the scores and dividing by the total number (N). The mathematical equation is:

$$\overline{X}=\frac{\Sigma(X_i)}{N}$$

The mean is the most sensitive measure of central tendency, which is to say that the mean will be affected by extreme scores in the distribution. As such, there are times when the median is reported for interval-ratio data rather than the mean, especially when there are a few extreme cases, such as with income in the U.S. To explain the difference of these measures, try the following exercise:

Get a ruler, some tape, and a few pennies. Try to balance the ruler on the tip of your index finger. If you are using a standard 12-inch ruler, the center should be around the six-inch mark. All things being equal, this point is both your mean and your median. It is the center of your distribution (0-12), or your median, and it is the arithmetic average of the scores (sum of 0 to 12 divided by number of scores, 13, = 78/13 = 6).

Now tape a penny to the 12-inch mark and <u>assume</u> that the value has doubled to 24. Now try to balance the ruler. You must move your "fulcrum", or the point around which all of the scores fall, up a bit, probably closer to the 7" mark, because of the added weight of the new value. If you calculate the mean, substituting the extreme score of 24 for 12, it is 6.9. The mean then is pulled up by the extreme score. The median, though, does not change. You still have 13 scores, which now range from 0 to 24; the exact center of your distribution is still six. The mean, though, is sensitive to extreme scores and should be used with some caution when a few extreme scores exist in your data.

This exercise also demonstrates the idea of skewness. If a distribution is **normal**, then the mean, the median, and the mode are all in about the same place: the exact center of the distribution. If a distribution has a few extremely high scores, then the mean is pulled upward, giving the distribution a **positive skew**. If a distribution has a few extremely **low** scores, though, the mean is pulled downward, giving the distribution a **negative skew**.

In the example with the ruler above, when the value 24 is substituted for the value of 12, the mean is pulled upward, giving the distribution a positive skew.

MULTIPLE-CHOICE QUESTIONS

1. Measures of central tendency are ______________________________.
   a. statistics that describe a population
   b. statistics that are used to find deciles
   c. statistics that summarize proportions
   d. statistics that report the most typical case

2. Choosing the correct measure of central tendency is based on ________________.
   a. sample size
   b. number of categories
   c. number of variables
   d. level of measurement

3. For nominal data, the correct measure of central tendency is _______________.
   a. median
   b. mode
   c. mean
   d. frequency

4. For ordinal data, the correct measure of central tendency is _______________.
   a. median
   b. mode
   c. mean
   d. frequency

5. For interval-ratio data, the correct measure of central tendency is _______________.
   a. median
   b. mode
   c. mean
   d. frequency

6. The median is __________________________.
   a. the most frequently occurring category
   b. the average case
   c. the number of cases
   d. the middle-most score

7. The mean is __________________________.
   a. the most frequently occurring category
   b. the average case
   c. the number of cases
   d. the middle-most score

8. The mode is __________________________.
   a. the most frequently occurring category
   b. the average case
   c. the number of cases
   d. the middle-most score

9. Percentiles are _________________ .
   a. the point below which a specific percentage of the cases fall
   b. the point above which a specific percentage of the cases fall
   c. the point which divides the distribution in half
   d. the point which divides the distribution in quarters

10. Quartiles are _________________ .
    a. the point below which a specific percentage of the cases fall
    b. the point above which a specific percentage of the cases fall
    c. the point which divides the distribution in half
    d. the point which divides the distribution in quarters

11. The term X_i means ____________________.
    a. the sum of the frequencies
    b. the sample size
    c. any category in a distribution
    d. any score in a distribution

12. For the variable of years married, which measure of central tendency would be most appropriate?
    a. mode
    b. median
    c. mean
    d. none of the above

13. For the variable of satisfaction with marriage, which measure of central tendency would be most appropriate?
   a. mode
   b. median
   c. mean
   d. none of the above

14. For the variable of favor or oppose inter-racial marriages, which measure of central tendency would be most appropriate?
   a. mode
   b. median
   c. mean
   d. none of the above

15. For the variable of age first married, which measure of central tendency would be most appropriate?
   a. mode
   b. median
   c. mean
   d. none of the above

WORK PROBLEMS

1. The data below give the ages of 20 students in a sociology statistics course. Find the mode, median, and mean. What does each tell us about the distribution? Is the distribution skewed? Which way? What happens to the distribution if the highest age is dropped?

18	19	20	21
19	18	20	20
20	19	18	21
20	20	20	21
19	20	19	37

2. For each of the frequency distribution tables below, find the most appropriate measure of central tendency:

a.

Number of Children in Family	f
0	20
1	10
2	56
3	32
4	10
5	2
6	0
7 or more	0
	130

b.

Social Class	f
Lower	5
Working	18
Middle	37
Upper-Middle	55
Upper	2
	117

c.

Political Party Affiliation	f
Democrat	543
Republican	469
Other	80
None	18
	1100

d.

Class Ranking	f
Freshmen	3
Sophomores	5
Juniors	15
Seniors	13
Graduates	1
	37

e.

Grades on Statistics Exam	f
100	2
95	12
85	15
75	18
65	15
55	5
	67

3. A-1 College polled their 1998 graduates recently to find out their incomes after graduating from their stellar programs. Below are the results of their poll.

A. Find the average income.

Incomes of Recent
Graduates of A-1 College

$ 12,500	$13,500
17,500	30,000
23,000	14,350
18,900	32,450
14,000	27,650
33,450	12,000
26,760	22,500
19,000	27,500
21,000	23,000
37,800	20,900

B. A-1 College did not include their most prominant alumni, Mike Rosoft, the computer genius whose income in 1998 was estimated to be $10,000,000. What would happen to the mean if his income was included? What would be a better measure to use if he were included?

DETAILED ANSWERS

1. Begin by organizing the data from lowest to highest. This makes it easier to see the distribution.

The mode, or most frequently occurring case, is 20. Therefore, there are more 20-year olds in this statistics class than any other age.

The median, the exact center of the distribution, is the average between the 10th and 11th cases (N/2 and N/2 + 1), which are 20 and 20. Therefore, the median is 20 years of age. Remember, you cannot find the median until the data has been ordered.

The mean, or the arithmetic average, is the sum of X_i divided by the total number of respondents (N), or 409/20, which is 20.45. Therefore, the average age of students in this statistics course is 20.45 years.

Because one student is a bit older than the rest, the average age is pulled upward, or has a positive skew.

If you were to drop that case, the mean age would be 19.6 years, which is closer to the mode and median, which are both still 20 years of age, making the distribution normal in shape.

2. For each distribution, the first question to ask yourself is what is the level of measurement of each variable? We cannot find the appropriate measure of central tendency until we know the level of measurement.

a. The level of measurement is interval-ratio, therefore we can use any of the three measures of central tendency, but the mean is usually the best measure. The mean is the sum of scores divided by N, or:

$$\overline{X} = \frac{(0x20)+(1x10)+(2x56)+(3x32)+(4x10)+(5x2)+(6x0)+(7x0)}{130}$$

$$= \frac{268}{130} = 2.06$$

Therefore, the average number of children per household is 2.06.

b. The level of measurement for this variable is ordinal, since we can rank-order the classes, but there is not a numerical value associated with each class. Therefore, we can use either the mode or the median, but the median is the best measure.

To find the median, we need to find the category associated with the 59th case (117/2=58.5). The easiest way to do so is to calculate the cumulative frequencies:

Social Class	f	cumulative f	
Lower	5	5	
Working	18	23	
Middle	37	60	<--------------59th case lies in this category
Upper-Middle	55	115	
Upper	2	117	
	117		

Therefore, the 59th case falls in the category "middle class". Note: the median is not the number 59, but the category in which the 59th case is in.

c. Political party affiliation is a nominal variable because you cannot rank-order the variable, but merely categorize it. Therefore, only the mode is the most appropriate measure of central tendency.

The mode is the most frequently occurring category which is Democrats, with 543 of the respondents falling into this category.

d. Class ranking is an ordinal variable since the categories can be rank-ordered from highest to lowest. Therefore either the mode or the median can be used, but the median is the better measure of central tendency.

Again, calculating the cumulative frequencies will make it easier to find the middle of the distribution:

Class Ranking	f	cumulative f	
Freshmen	3	3	
Sophomores	5	8	
Juniors	15	23	<-------- 19th case falls in this category
Seniors	13	36	
Graduates	1	37	
	37		

The median is the category associated with the 19th case (37/2 = 18.5), which is the category "Juniors". Therefore, half of the distribution falls both above and below the 19th case in the category of Juniors.

e. The level of measurement of the grades is interval-ratio, which means you can use any of the three measures of central tendency, but the best measure is usually the mean.

To find the mean, sum all of the scores and divide by the sample size, N.

Because the scores are in a frequency distribution, the easiest way to do this would be to multiple each value by its frequency, and then add them together, and divide by N:

$$\overline{X} = \frac{(100x2)+(95x12)+(85x15)+(75x18)+(65x15)+(55x2)}{67}$$

$$= \frac{5{,}215}{67} = 77.84$$

Therefore, the average grade in this class is 77.8.

3A. Income is an interval-ratio variable, so the mean can be used. Often, though, there is such a wide range of incomes that the median is used. In order to determine which measure is best, find both the median and the mean.

To do so, organize the data from lowest to highest. Find the median, which is the <u>average</u> of the 10th and 11th cases or:

$$\frac{(21{,}000) + (22{,}000)}{2} = 21{,}700$$

The middle-most income, then, is \$ 21,700.

The mean is the sum of scores divided by N, or:

$$\frac{44{,}7760}{20} = 22{,}388$$

Therefore, the average income is \$ 22,388. The mean is a bit higher than the median which indicates a slight positive skew to the distribution. Although the difference is not extreme, it is good idea to report both the median and the mean.

B. If we include Mike Rosoft's income of $10,000,000 the mean is:

$$\frac{10,447,760}{21} = 497,512.38$$

Therefore, the average income of the 1998 graduating class would be close to $500,000. If you look at the distribution, though, that does not represent the sample in the least! Therefore the median would be a better measure of central tendency.

CHAPTER FOUR
MEASURES OF DISPERSION

Measures of dispersion are another way of **describing** a distribution of scores on some given variable. Like measures of central tendencies, measures of dispersion give the reader a sense of what the sample looks like, but with a bit more information. We know what the average score looks like with either a mode, median, or mean, but with these new measures we can go a step further and ask "How scattered are the data around the average score?"

If we have a sample of students from a statistics class and found that the average (mean) age is 20, we probably also want to know where the rest of the sample falls. If it ends up that everyone in the sample is 20 years old, then there is virtually no dispersion in our sample. Everyone is the same age, therefore there is no scattering of the ages across a distribution. More likely, though, the average age will only represent the majority of the class, but there will be many others who are not 20 years old. The measure of dispersion will find out how different the rest of these students are in age from those who are 20 years old.

Let's say we have several people in our sample who are not 20 years old, but instead range in age from 17 to 47 years of age. Although we found the average, or mean, age to be 20, we also know that there is quite a bit of disparity between the ages. We would expect then a high degree of dispersion in this sample with regard to age.

Like measures of central tendency, there are different measures of dispersion based on the level of measurement of the variable in question. In the example above, we were talking about age which is an interval-ratio variable. The appropriate measure of central tendency then is the mean; the appropriate measure of dispersion is **standard deviation**. Likewise for nominal data, we use the **index of qualitative variation (IQV)**, and for ordinal data we use **IQV,** the **range (R)**, or the **interquartile range (Q)**. A simple way to remember these is with the chart below:

LEVEL OF MEASUREMENT	MEASURE OF CENTRAL TENDENCY	MEASURE OF DISPERSION
NOMINAL	MODE	IQV
ORDINAL	MEDIAN	IQV, R, Q
INTERVAL-RATIO	MEAN	STANDARD DEVIATION

The **index of qualitative variation** (or **IQV**) is the only measure of dispersion which can be used on nominal data, and measures how much variation there is in a sample on a nominal-level variable. In Table 4.1 of your textbook, there is a good example how the IQV is used. Data for the racial composition of three neighborhoods is given. In the first neighborhood, everyone is white: there is no variation in the racial composition of this neighborhood. In the second neighborhood, the majority is white, but there is a bit of variety as well. In the third neighborhood, the racial composition is spread evenly across the three racial groups:

therefore, there is maximum variation in this neighborhood.

Now think about your own neighborhood. Is the majority of the population white? Is the majority black or hispanic? Perhaps there is no one majority; what can you tell about the amount of variation of your neighborhood?

The index of qualitative variation enables us to measure this variation by looking at the frequencies in each of the categories, and ranges between 0 and 1. In the first neighborhood, there were 90 people, all of whom are white. There is no variation in this neighborhood, therefore when we calculate the IQV measure, it is 0, which indicates no variation.

The equation for IQV is:

$$IQV = \frac{k(N^2 - \sum f^2)}{N^2(k-1)}$$

where k = the number of categories
N = the number of cases
$\sum f^2$ = the sum of squared frequencies

The number of categories (k) is simply the number of groups in your distribution. In the example above, we have three groups or categories: white, black, and other (k=3). If we were analyzing the variable gender, we would have two categories: male and female (k=2). The $\sum f^2$ is simply the squared frequencies in each category, summed. In the first neighborhood, we had 90 whites, therefore the squared frequencies is 8100 (90^2=8100). Since there were zero frequencies in the other two categories, the sum of the squared frequency is only 8100. See? Simple!

The measures of dispersion for ordinal variables are not as simple, primarily because we are trying to squeeze our measure to fit this type of data. Ordinal data, remember, are data which are organized into some hierarchy, from highest to lowest, but exact distances between two scores are unknown. Thus, ordinal data are usually categories such as freshman, sophomore, junior, senior; or scaled items such as Strongly Agree, Agree, Disagree, Strongly Disagree, not numbers. If we were to find the appropriate measure of central tendency, we find the median: the middle-most score and its corresponding category.

`For measures of dispersion, then, we can talk about the **range** of scores or categories, such as the scale ranges from Strongly Agree to Strongly Disagree, or from Freshman to Senior. We can also find the **interquartile range (Q)**, which locates where the middle 50% of the cases fall in the distribution. If we are using true ordinal data, we can find the one category or multiple categories where 50% of the distribution falls. Remember, if our data is in true categories, with no zero point, then the quartiles are these categories, not the frequencies associated with each. Because of the "stickiness" of ordinal data, it is difficult to use this measure, and most often the range is used to measure for ordinal variables.

The most important, and most widely used, measure of dispersion is the **standard deviation** (or **s**), which can only be used with interval-ratio data. Think of the term "deviations." We all have probably met or know someone whom we consider a deviant--someone who does not quite fit our definition of normal. Standard deviations are interested in those people: the ones who are not the average, those whom are not 20 years old, etc. Thus, we can find the most typical, or average case, and then find out how much our sample deviates from that case.

Think of all the students in your statistics course. If the average age is 20, how far from the mean are you? How far from the mean is the person next to you? If you subtract your age from the mean age (20), is there a difference or does it equal zero?

The **definitional equation** for standard deviation is:

$$s = \sqrt{\frac{(\Sigma X_i - \overline{X})^2}{N}}$$

Each value is subtracted from the mean, squared and summed, and divided by the sample size. The square-root of this value indicates the amount of variation there is in the sample. If everyone is basically the same age, there will be very little variation; if there are a lot of different ages, there will be a lot of variation. Two other formulas exist that are easier to use to calculate the standard deviation:

$$s = \sqrt{\frac{\Sigma X^2_i}{N} - \overline{X^2}}$$

or

$$s = \frac{1}{N}\sqrt{N\Sigma X^2_i - (\Sigma X_i)^2}$$

Unlike the IQV measure, the standard deviation does not have a set range. It can be as low as zero (meaning no deviations from the mean--or more simply put: everyone is exactly the same), and as large as infinity. The larger the value, the more dispersed the scores are around the mean. A good rule of thumb is that a standard deviation of approximately half the size of the mean represents a reasonable "normal" curve (which will be discussed in greater detail in Chapter 5).

Putting the standard deviation into words will be easier once you understand the concept of the "normal curve" (Chapter 5), but at this point you can speak of the dispersion as plus or minus one standard deviation away from the mean, or if the mean age is 20, and the standard deviation is three, we can say that the average age is 20 years, plus or minus three years.

MULTIPLE-CHOICE QUESTIONS

1. Deviations are:
   a. the amount of variety in a distribution of scores
   b. the highest score minus the lowest score
   c. the distance between the scores and the mean
   d. changes in the quality of the scores

2. Dispersion is:
   a. the amount of variety in a distribution of scores
   b. the highest score minus the lowest score
   c. the distance between the scores and the mean
   d. changes in the quantity of the scores

3. IQV is the acronym for:
   a. Interrelate Qualitative Variance
   b. InterQuartile Variance
   c. Index of Qualitative Variation
   d. Inner Quality of Variety

4. The IQV measure is the only measure for ___________ level of measurement.
   a. nominal
   b. ordinal
   c. interval-ratio
   d. all of the above

5. When using the IQV measure, k is:
   a. sample size
   b. number of cases in each category
   c. number of categories
   d. sum of frequencies, squared

6. The IQV measure of dispersion ranges from ____ to _____.
   a. -1 to +1
   b. 0 to 1
   c. 0 to infinity
   d. 0 to 10

7. The range is:
a. the highest score minus the lowest score
b. the distance between the scores and the mean
c. a measure of dispersion for nominal data
d. affected by the sample size

8. Standard deviation(s) is (are):
a. a measure of dispersion for interval-ratio variables
b. the square root of the squared deviations of the scores around the mean divided by N
c. a measure of dispersion for a sample of a population
d. all of the above

9. Variance is:
a. the squared deviations of the scores around the mean divided by N
b. used primarily for ordinal data
c. compares the number of cases with the number of categories
d. all of the above

10. Standard deviations range from ______ to _______.
a. -1 to +1
b. 0 to 1
c. 0 to infinity
d. 0 to 10

11. A standard deviation of zero indicates:
a. nothing
b. complete variety in the distribution
c. the mean and standard deviation are the same
d. no variety in the distribution

12. The definitional formula of the standard deviation is the ____________________.
a. square root of the variance
b. sum of the means across categories
c. mean minus the highest score
d. k minus the sum of the means, squared

13. The computational formulas of the standard deviation are often used because ________________________.
a. they are more accurate
b. algebraically, they yield a better measure of dispersion
c. they base their calculations on the number of categories
d. they are easier to use

14. If you have an average age of 35 with a standard deviation of 37 years, what might you conclude about the distribution?
   a. The scores are clustered around the mean.
   b. The average age is 35, plus or minus two years.
   c. There is very little variation in the sample on this variable.
   d. There is a lot of variation in the sample on this variable.

15. If the average score on a statistics exam was 76, and the standard deviation was 7, what can you conclude about the distribution?
   a. The scores are clustered around the mean.
   b. The average score on the exam was 76, plus or minus 7 points.
   c. There is very little variation in the sample on this variable.
   d. All of the above

WORK PROBLEMS

For each of the frequency distribution tables below, find the most appropriate measure of dispersion:

1.

Number of Children in Family	f
0	20
1	10
2	56
3	32
4	10
5	2
6	0
7 or more	0
	130

2.

Social Class	f
Lower	5
Working	18
Middle	37
Upper-Middle	55
Upper	2
	117

3.

Political Party Affiliation	f
Democrat	543
Republican	469
Other	80
None	18
	1100

4.

Class Ranking	f
Freshmen	3
Sophomores	5
Juniors	15
Seniors	13
Graduates	1
	37

5.

Grades on Statistics Exam	f
100	2
95	12
85	15
75	18
65	15
55	5
	67

6.

Incomes of Recent College Graduates

$ 12,500	13,500	21,000
17,500	30,000	37,800
23,000	14,350	14,000
18,900	32,450	33,450
26,760	22,500	23,000
19,000	27,500	20,900
27,650	12,000	

7.

Age	f
18	3
19	5
20	8
21	3
37	1
	20

8. Religious Preference

Religious Preference	f
Protestant	24
Catholic	37
Jewish	18
Other	7
None	3
	82

DETAILED ANSWERS

1. Since the data are interval-ratio, you need to find the standard deviation. One way to do it would be to square each case separately, then sum them. This would be tedious, so I suggest squaring the value for each category, then multiplying that by the number of cases, then summing these together.

Number of Children	Value^2	x	f		
0	0		20	=	0
1	1		10	=	10
2	4		56	=	224
3	9		32	=	288
4	16		10	=	160
5	25		2	=	50
6	36		0	=	0
7 or more	49		0	=	0
				$\sum X_i^2$ =	732

$\overline{X}=2.06$

$$s=\sqrt{\frac{\Sigma X_i^2}{N}-\overline{X^2}}=\sqrt{\frac{732}{130}-(2.06)^2}=1.18$$

Therefore, the sample has an average of 2.06 children, plus or minus 1.18 children.

2. This variable is ordinal, and as such, it is a sticky variable to find the measure of dispersion. It is possible to talk about the range of the values (the range is between upper class and lower class), but that doesn't make much sense. We could for the sake of ease compute the nominal measure of dispersion, IQV, or we could compute the interquartile range (Q). I'd opt for the easier method, simply because it will be a decent description of the dispersion of the data, and will be easier to interpret.

$$IQV = \frac{k(N^2-\sum f^2)}{N^2(k-1)} = \frac{5(13689-4747)}{13689(4)} = 0.82$$

Since IQV ranges between 0 and 1, 0.82 indicates quite a bit of variation in our sample.

If you wanted to do Q, Q1=29th case; Q3=88th case and Q=59

3. Political party affiliation is nominal data, so we should use the IQV:

$$IQV = \frac{k(N^2-\sum f^2)}{N^2(k-1)} = \frac{4(1100^2-521534)}{1100^2(3)} = 0.76$$

Again, this indicates quite a bit of variation in our sample.

4. Class ranking is ordinal level data, but, again, I'm going for the measure that is easiest to interpret (I do this only for measures of dispersion with ordinal data!), which would be IQV:

$$IQV = \frac{k(N^2-\sum f^2)}{N^2(k-1)} = \frac{5(1369-429)}{1369(4)} = 0.86$$

There is a lot of variation in our sample for the variable class ranking.

5. This variable is interval-ratio, so we can compute the standard deviation, which is:
$\sum X_i^2=416425$

$\bar{X}=77.8$

$$s=\sqrt{\frac{\sum X_i^2}{N}-\bar{X}^2}=\sqrt{\frac{416425}{67}-(77.8)^2}=12.75$$

Therefore, the average grade on the statistics exam was 77.8, with a standard deviation of 12.75.

6. Here again is interval-ratio data, so we should compute the standard deviation. If we square the value of each score and sum all those values, we are going to get a very large, very scary number. Thus, here's another quick trick: divide each score by 1000, work with these numbers, then multiply be 1000 at the end of the problem. You should get numbers like these:
$\sum X_i^2=11088.558$

$\bar{X}=22.388$ (x 1000 = \$22,388)

$$s=\sqrt{\frac{\sum X_i^2}{N}-\overline{X}^2}=\sqrt{\frac{11088.558}{20}-(22.388)^2}=7.2941988$$

7.2941988 x 1000 = 7294.2

Therefore, the average income of recent college graduates is $22,388, with a standard deviation of $7,294.

7. Again, we need to compute the standard deviation for interval-ratio data:

$\sum X_i^2=8669$

$\overline{X}=20.45$

$$s=\sqrt{\frac{\sum X_i^2}{N}-\overline{X}^2}=\sqrt{\frac{8669}{20}-(20.45)^2}=3.91$$

Therefore, the average age is 20.45 years, plus or minus 3.9 years.

8. Since this is nominal data, we must use the IQV measure:

$$IQV = \frac{k(N^2-\sum f^2)}{N^2\,(k-1)} = \frac{5(6724-2327)}{6724(4)} = 0.82$$

Again, there is quite a bit of variation in this sample on the variable religious preference.

CHAPTER FIVE
THE NORMAL CURVE

In chapter two you learned about histograms, which are graphs used to graphically display the frequencies of interval-ratio variables. In histograms, the frequencies are displayed with contiguous bars, each one marking the frequency for one category of the variable in question. Now, consider drawing a curve that copies the shape of that distribution. If that curve is bell-shaped, with one mode (unimodal), and the median and mean are about the same place as the mode, we could conclude that this variable has a **normal distribution**, and is shaped by a **normal curve**.

Theoretically, a normal curve should be perfectly symmetrical, with 50% of the cases falling on each side of the highest point, which theoretically is the mean, median, and mode. It would be very rare indeed for a distribution of any variable to exactly model the theoretical normal curve, but it is important for the assumption of normality. With this assumption met, we can describe our distribution, find the area above and below the mean, as well as between scores, and make predictions.

When describing a distribution, we previously used the standard deviation. With the assumption of the normal curve (normality) met, we can then describe our distribution in terms of the percentage of respondents above and below the mean. In the case of a normal curve, we can divide the area under the curve into segments, each one representing one standard deviation. Therefore, we can find the area under the normal curve which corresponds to the area of one standard deviation away from the mean (plus or minus). Lucky for us, this area has already been found: 68.26% of the respondents will **always** fall plus and minus one standard deviation from the mean. In English this means that 68.26% of the respondents are within the area of one standard deviation **below** the mean and one standard deviation **above** the mean.

We also are given the areas two and three standard deviations away from the mean. The area under the normal curve two standard deviations away from the mean encompasses 95.44% of the sample; the area under the normal curve three standard deviations away from the mean encompasses 99.72% of the sample. These will always be the same for any normal curve (therefore, I would suggest memorizing them!)

If 99.72% of all the scores lie within three standard deviations from the mean, then only a very small percentage (0.28%) fall beyond three standard deviations from the mean. Also, if 68.26% of the sample falls within one standard deviation of the mean, then we also know that the majority is in that area. Thus, the majority of the respondents cluster around the mean, giving the normal curve its shape.

We can also find the total area below a score, or even the total number of people below a given score. If we know that the average (mean) age in a statistics course is 20 years old, with a standard deviation of three years, and you are 23 years old, what percentage of the rest of the class is younger than you? What percentage of the class is older than you? How far above the average age are you? We can answer all of these questions by converting our raw scores into **Z scores**.

Z scores are standardized scores in a distribution. Basically it works like this: in the U.S. we use the Fahrenheit scale to measure temperature, which has a freezing point of 32°.

In other parts of the world, the Celsius scale is used which has a freezing point of zero. If we were to change our description of the temperature outside today from Fahrenheit to Celsius, we are not changing the temperature, but only the **scale** on which we are measuring it. Z scores work the same way. We are not changing the scores, as much as changing the scale in which we are measuring the variable.

As such, Z scores will always have a mean of zero and a standard deviation of one. Thus in the example from the statistics class, if in the raw scores there was a mean age of 20 years old, and a standard deviation of 2.5 years, when we convert these scores to Z scores, the mean age is now zero and the standard deviation is one. To convert any score into a Z score, we use the following equation:

$$Z=\frac{X_i-X}{s}$$

This equation, then, converts any score into a Z score. Thus if you wanted to convert your own age into a Z score, you would plug your age into X_i and complete the equation. If you were 23 years old, the equation would be:

$$Z=\frac{23-20}{2.5}=\frac{3}{2.5}=+1.2$$

Therefore, if you are 23 years old, in a class with the average age of 20 years old and a standard deviation of 2.5 years, you are 1.2 standard deviation units **above** the mean. Well, so what?, you might say. So what, indeed! Now the fun begins. If you look at Appendix A in the back of your textbook, you will see several columns. This is the area under the normal curve.

Find the Z score (column a) of 1.20. In column b, the area between the mean and Z, is 0.3849, or 38.49% of the respondents fall between the mean (20 years old) and your Z score of 1.20 (23 years of age). To find the area above your score (or those whom are older than 23), go to column c: 0.1151 or 11.51% of the class is older than 23.

Z scores can also be used to find the total proportion or percentage of scores which fall below a certain score. Remember that 50% of all scores fall above and below the mean. As such, we know that 50% will always be below the mean, and 50% will always be above the mean. If we have a positive Z score, then we can find the area between the mean and Z, and **add** 50% to find the **total area below** that score. Thus, if you are 23 years old in the above example, 38.49% of the sample falls between your age and the mean, and 88.49% of the sample is younger than you (50% + 33.49% = 88.49%).

Similarly, if you want to find the area **between** two scores, let's say your age of 23 and someone who is 18 years old, you need to convert both scores to Z scores and add the corresponding area. We know the area between the mean and 23 is 38.49%; the area between the mean and 18 years of age is:

$$Z=\frac{18-20}{2.5}=\frac{-2}{2.5}=-0.8$$

Someone who is 18 years old in this class is 0.8 standard deviation units **below** (negative number) the mean. The area between the mean and the age of 18, then (column b) is 0.2881; or 28.81% of the sample falls between the ages of 18 and 20.

We can now find the area between these two scores by adding the areas between each score and the mean:

28.81% + 38.49% = 67.3%

Therefore, 67.3% of the sample falls between the ages of 18 and 23.

We can also make predictions with Z scores. Since the area between the mean and Z is in proportions, we can consider, for example, the likelihood that a randomly selected respondent will be between the age of 20 and 22. First we need to find the Z score:

$$Z = \frac{X_i - \overline{X}}{s} = \frac{22-20}{2.5} = +0.80$$

The corresponding area for a Z score of 0.80 (column b) is 0.2881. Therefore, the likelihood that a randomly selected student from our sample will be between the ages of 20 and 22 is 0.2282. Since proportions range from 0 to 1, zero meaning no chance at all, and 1 meaning likely, there is a 0.2 chance of randomly selecting a person who is between the ages of 20 and 22.

Remember: Z scores are to be used for interval-ratio data only. You want to standardize scores to find the area below a normal curve, and the corresponding number or percentage of cases either above or below the mean, between the mean and a particular score, or the total percentage (or number or proportion) of scores falling above or below a given score.

MULTIPLE-CHOICE QUESTIONS

1. Z scores can be computed for only __________ type of data.
   a. nominal
   b. ordinal
   c. interval-ratio
   d. all of the above

2. The normal curve is:
   a. a theoretical distribution of scores
   b. unimodal
   c. bell-shaped
   d. all of the above

3. A normal curve will have a mean of ______ and a standard deviation of _____.
   a. 0, 0
   b. 0, 1
   c. 1, 0
   d. 1, 1

4. Z scores are:
   a. unidiminsional scores
   b. nondenominational scores
   c. rounded scores
   d. standardized scores

5. In the Z equation, the mean is subtracted from:
   a. an unknown score
   b. a given score
   c. itself
   d. all other scores

6. In the Z equation, the denominator is:
   a. the standard deviation
   b. one
   c. both a and b
   d. none of the above

7. To find the area between a score and the mean, we use the _____ column in Appendix A.
   a. (a)
   b. (b)
   c. (c)
   d. (d)

8. To find the area below a positive Z score, you need to:
   a. subtract the mean from the Z score
   b. find the corresponding Z
   c. add 50% to the corresponding Z
   d. subtract 50% from the corresponding Z

9. To find the area below a negative Z score, you need to:
   a. find the corresponding Z
   b. find the area beyond the Z
   c. find the area between the mean and Z
   d. add 50% to the corresponding Z

10. To find the area above a negative Z score, you need to:
   a. subtract the mean from the Z score
   b. find the corresponding Z
   c. add 50% to the corresponding Z
   d. subtract 50% from the corresponding Z

11. To find the area between two scores, you need to:
   a. find the corresponding Z scores
   b. find the area between each Z score and the mean
   c. add the two corresponding Z scores
   d. all of the above

12. Using Appendix A in your book, what is the area between the mean and a Z of 1.71?
   a. 50%
   b. 4.36%
   c. 1.71%
   d. 45.64%

13. Using Appendix A in your book, what is the area beyond a Z of 3.43?
   a. 50%
   b. 3.43%
   c. 0.03%
   d. 49.97%

14. Using Appendix A in your book, what is the total area above the mean?
   a. 50%
   b. 1%
   c. 0.01%
   d. 45.62%

15. Using Appendix A in your book, if the area between the mean and Z is 11.79%, what is the corresponding Z?
   a. 0.1179
   b. 0.3821
   c. 0.30
   d. 0

WORK PROBLEMS

1. A researcher has data on starting salaries of recent social science graduates. The average salary is $27,654, with a standard deviation of $4,569. Assuming a normal distribution, find the corresponding Z scores for the following salaries.

   a. $35,456
   b. $23,875
   c. $27,450
   d. $25,000
   e. $42,750

2. Using the same data, what percentage of the sample falls below the salary of $30,000?

3. Bill and Jeff are graduating in a few months and both have found employment. Bill found a job that he loves, but the pay is below average ($25,000); Jeff is so-so on his new job, but since the pay is high, he decided to take it ($33,500). What percentage of their class will be receiving salaries between theirs? Use the data from Question 1 above.

4. The average years of education in a sample of business executives is 17.2 years, plus or minus 2.3 years. A person graduating with a bachelors (16 years of education) is hoping to be a business executive someday. How far above or below the average years of education is he/she?

5. Sue is advanced for her age, and graduated high school at the age of 12. She is now a sophomore in college at the ripe old age of 14, but realizes she is quite a bit younger than her cohort, whose average age is 19.8, with a standard deviation of 1.3. How much younger than the rest of the class is she? What percentage of her class is younger than she?

6. Poor Sue, being the bright young woman she is, realizes that the majority of the class is older than she. What percentage of the class is between 18.5 years and 21.1 years of age?

7. If the average grade on the first statistics exam is 76% with a standard deviation of 5.9%, what can you say about the distribution?

8. Sue, whom we met above, earned a 92% on the first exam. How far above the mean is her score? What percentage of the class had a higher score? What percentage of the class had a lower score?

9. Jeff, who teased Sue about being so young, earned a 67% on the first exam. How far from the mean was his score? What percentage of scores fell between Jeff and Sue's scores?

10. The average number of children a sample of Americans have had is 2.3, with a standard deviation of 1.7. What percentage of the sample have had more than three children?

11. What is the probability that a randomly selected person from our sample will have between 1.9 and 2.3 children?

12. The employees at Davis and Davis live an average of 10.3 miles from the office (s = 5.3 miles). The CEO is considering moving the office to a new location further from the city center. The new office space will still be an average of 10.3 mile commute for the employees (± 5.3), but who will benefit the most? How far away from the average commute is each person's commute for both the former and new offices?

	Former office	New office
Donna	12.3	9.5
Jim	10.8	14.8
Judith	3.0	6.4
Jose	14.5	11.5
Robert	2.5	4.7

13. The average SAT score for entering freshmen at a large university is 1100, with a standard deviation of 250. Bill knows he needs to be within one standard deviation of the mean to get accepted. What is the lowest score he can earn on the SATs?

14. Ruthy recently earned a 1230 on the SATs. She knows she's above the minimum requirement, but she wants to know what percent of the entering freshmen fell below her score and what percent fell above. Help her out.

15. Carl has eyes for Ruthy. He knows she is very bright and wants to make a good impression on her. He recently found out that her score was 1230 on the SATs, while his were only 1150. How far apart are their scores?

DETAILED ANSWERS

1. a. Z = +1.71, which means that this salary is 45.64% above the average salary.

b. Z = -0.83, which means that this salary is 29.67% below the average salary, or that about 30% of the sample have salaries between the $23,875 and the average, $27,645.

c. Z = -0.045, or about -0.05, which has a corresponding area under the normal curve of 0.0199. Thus, this salary is 1.99% below the average salary.

d. Z = -0.58, which means that 21.90% of the sample have salaries between the $25,000 and the mean.

e. Z = + 3.30. This salary is 49.95% higher than the average salary. Only 5% of the sample is likely to have salaries higher than $42,750.

2. Z = 0.51 The area between the average salary and a salary of $30,000 is 19.50%. 69.50% of the sample falls below a starting salary of $30,000.

3. Bill has a Z score of -0.58, Jeff has a Z score of +1.28. The corresponding areas are 21.90% and 39.97%, respectively. The percentage of respondents who have salaries between them is 61.87%

4. He/she is 0.52 standard deviations **below** the average years of education. Therefore, 19.85% of the sample is between this score and the mean.

5. Sue is 4.46 standard deviation units below the average age in her class. Less than 0.01% of the class is younger than she.

6. If you add and subtract one standard deviation away from the mean, you will get 18.5 and 21.1. Since we know that 68.26% of a sample falls within one standard deviation of the mean, we know that 68.26% of the class is between these two ages.

7. As in Question 6, if we add and subtract the standard deviation from the mean, we can estimate the distribution. Here, 68.26% of the sample is one standard deviation below the mean and one standard deviation above the mean, or between the exam scores of 70.1 and 81.9.

8. A grade of 92% has a Z score of +2.71, which indicates that Sue is 49.66% above the average grade, and 99.66% above the entire class. Thus, only 0.34% of the class scored higher than Sue on the exam.

9. Jeff is 1.53 standard deviation units below the mean, which means that he is 43.57% below the average score. Since Sue's score was 49.66% above the mean, the area between Jeff and Sue's scores is 93.23%.

10. The Z score associated with three children is 0.41; the area beyond 0.41 (column c) is 0.3409. Therefore, 34.09% of the sample have had more than three children.

11. If you subtract 1.9 from 2.3, and divide by the standard deviation, you get a Z score of -0.24. The corresponding area between the mean and score (column b) is 0.0948. Therefore, there is a one in 9.5 chance that a randomly selected person from the sample will have between 1.9 and 2.3 children.

12.

	Former office	Z	New office	Z
Donna	12.3	0.38	9.5	-0.15
Jim	10.8	0.09	14.8	0.85
Judith	3.0	-1.38	6.4	-0.74
Jose	14.5	0.79	11.5	0.23
Robert	2.5	-1.47	4.7	-1.06

Donna seems to have cut down her commute time the most, from being 0.38 standard deviations above the mean, to being 0.15 standard deviation units below the mean. Robert seems to still be the closest to the office, though.

13. The lowest SAT score Bill can earn is 850, which is 1100 minus one standard deviation (250).

14.

$$Z = \frac{1230 - 1100}{250} = 0.52$$

Ruthy is 0.52 standard deviation units above the mean, which indicates that 19.85% of the sample is between her score and the mean, with 30.15% of the sample having a higher score. Therefore, 69.85% of the sample fell below her score.

15. Carl has a Z score of 0.20, which indicates that he is 7.93% above the mean. If Ruthy is 19.85% above the mean, the area between them is 11.92%. Close enough for compatibility, Carl decides.

CHAPTER SIX
INTRODUCTION TO INFERENTIAL STATISTICS
Sampling and the Sampling Distribution

In the first five chapters of your text, you learned **descriptive statistics**: ways to describe the distribution of your sample on some given variable. In the next section, we are moving onto **inferential statistics**: ways in which we can infer our sample results to the rest of the population. Two important and very distinct ideas here are (1) samples, and (2) populations. Populations are the entire body of who or what you are studying. If you are interested in women of childbearing age in the U.S, then your population is all women between the ages of 13 and 50 in the United States. Since it is virtually impossible to reach all of these women, you need to sample this population. That is, you need to randomly select a subset of the population which will represent all women in the U.S. between the ages of 13 and 50.

In order to do so, you must use a probability sampling technique, also called "random sampling", such as **simple random sampling**, **systematic sampling, stratified sampling**, or **cluster sampling**. Each of these is discussed in greater detail in your textbook, but the basic idea behind all of this is that every single person in your population has an equal probability of being selected for your sample, thus assuring that your sample is representative of the entire population. For example, let's say that you were interested in two variables, height and shoe size, and decided to do a study on the relationship between the two variables. In order to find a quick sample, you decided to walk to the gym where the men's basketball team is just finishing practice. As each player leaves the court, you ask each one his height and his shoe size. You then take this data home and find that everyone is very tall and has very large feet. In fact, you go so far as to conclude that height and shoe size are related: if you have large feet, you are tall and if you are tall, you have large feet.

What's wrong with your data? Well, your sample is not representative of the whole population. In fact, it does not even represent all basketball players. It is only using one population: the basketball players in that particular gym at that time. Because of the non-randomness of your sample, you cannot infer your results to anyone besides the basketball players in the gym.

Thus, social scientists ensure that they can infer their results by utilizing probability sampling techniques. Describing a sample is fine and good, and in fact is an important first step in analyzing data, but the main point in social science research is to make a statement about the *entire* population, not just the sample. Therefore, inferential statistics are a mainstay to statistical analysis.

In order to make these inferences, social scientists make use of something called the **sampling distribution**, which is a theoretical notion that if you could take an infinite number of samples, of let's say 100 respondents each (N=100), and you plotted each sample's mean of height, the sampling distribution of means would represent a normal curve. That is, we are making the assumption, with the use of this theoretical sampling distribution, that if enough samples are collected, the distribution of these sample means will be shaped normally.

Stated in another way, we are saying that we know that in theory that an infinite number of samples of all of the same size will produce a distribution of means that looks like a normal distribution. Because we know this to be true, then, we can assume that if we take only one sample of that same size, it will be within the range of the normal distribution and thus be representative of the population.

But why in the world would anyone ever care about such things, you ask yourself. Because without this assumption, we cannot make inferences about the population. We need this theoretical framework in order to say that our sample results <u>can</u> be inferred to the population, which, as we stated before, is what we live for.

MULTIPLE-CHOICE QUESTIONS

1. Parameters are:
   a. statistical methods of inferential statistics
   b. the distance between descriptive and inferential statistics
   c. sampling techniques for inferential statistics
   d. generalized population characteristics

2. When we talk of a sample average, we call it a mean, for a population average we call it:
   a. standard mean
   b. sigma
   c. parameter
   d. mu

3. Probability samples are also called ___________ samples.
   a. inferential
   b. continuous
   c. random
   d. population

4. EPSEM is the acronym for________________ ____________________.
   a. Equal Probability of SElecting Method
   b. Equal Probability of Service Employment Method
   c. Earnest Plying of Selected Entry Method
   d. Every Person Serves Equally Method

5. A sample is representative if it ____________.
   a. falls within the sampling distribution
   b. is greater than 1000
   c. reproduces the major characteristic of the population
   d. all of the above

6. A simple random sample is a method in which ____________________.
   a. every case has an equal chance of being chosen
   b. every other case is chosen for the sample
   c. the sample predicts the population
   d. the population is used to estimate the parameter

7. A clustering sample is a method in which ____________ units are randomly selected and ____________ case(s) within each is/are tested.
   a. demographical, every kth
   b. population, minimum
   c. sampling, every 100th
   d. geographical, all

8. Stratified sampling selects a sample from:
   a. generalized tracts in the population
   b. sublists of the population
   c. random numbers table
   d. demographic transition table

9. The central concept to inferential statistics is:
   a. EPSEM
   b. stratified sampling techniques
   c. sampling distribution
   d. population parameters

10. The three distributions involved in every application of inferential statistics are:
    a. sample proportion, population proportion, sampling mean
    b. random sampling, sample distribution, sampling probability
    c. sample distribution, population distribution, sampling distribution
    d. EPSEM, sample means, sampling distribution

11. The sample distribution is:
    a. known
    b. theoretical
    c. random
    d. isomorphic

12. The population distribution is:
    a. known
    b. theoretical
    c. random
    d. empirical

13. The sampling distribution is:
   a. known
   b. theoretical
   c. empirical
   d. random

14. The mean of the sampling distribution is represented by the letter/symbol of:
   a. μ_x
   b. μ
   c. X
   d. σ

15. _________ letters/symbols are used to denote population values.
   a. Capital
   b. Lower case
   c. Roman numeral
   d. Greek

CHAPTER SEVEN
ESTIMATION PROCEDURES

Chapter seven introduces you to two types of estimation procedures: point estimates and confidence intervals. For both types of estimation procedures, we are attempting to **estimate** the *population* values by making use of our *sample* values. In order to do so, our estimators must be both unbiased (the mean of the sampling distribution is equal to the population mean) and efficient (the extent to which the sampling distribution is clustered around the mean) (Healey, 1999: 154). Basically this means that we are assuming our estimator to be unbiased, and know that as our sample size increases, so too does the efficiency of our estimator.

Point estimates use a single-sample estimate as an estimator of the population, meaning that point estimates find a sample mean and use that as an estimator of the population mean. **Confidence intervals** use a single-sample estimate to find a range in which we would estimate the population value to fall within. This range then is the confidence interval, in which we are fairly confident that the population value will fall between two scores.

Here's a nice example of how confidence intervals work (adapted from Blalock, 1979[1]): Let's say you have a block of wood and you weigh it and it weighs 8 ounces. The weight of wood, being the breathable material that it is, though, is not constant. So you weigh it a second time and it weighs 7.5 ounces. A third weighing indicates that it is 8.9 ounces; a fourth is 9.3; and again at 7.8; and so on. After you weigh this wood several more times, you can get an average weight of the piece of wood, let's say 8.3 ounces, plus or minus 0.76 ounces. Therefore you are admitting that you don't know exactly how much this piece of wood weighs, but you're fairly confident that its weight is in this range, or confidence interval. Remember, this is just an example. Confidence intervals are used to estimate a range of **population values** from a **sample value**.

How confident we are that this range truly does contain the population value depends largely on what we set our error to. We decide beforehand how often we are going to be right. Do we want to be correct 50% of the time? No, that's too low. How about 75% of the time? Well, that means we're going to be wrong 25% of the time--still too low. Ok, how about 95% of the time. Well, that seems pretty reasonable. Let's say we're going to be correct about estimating the population value 95% of the time. That means we're going to be wrong only 5% of the time. Pretty good odds, huh! Making this decision, the probability of being wrong, is called choosing an **alpha level**. Here, we decided to set our alpha level at 0.05, or 5%, which means that we are going to be wrong only 5% of the time. We could set our alpha at any level, but 0.05 is used most often in the social sciences.

Ok, so we've decided to estimate the population value from our sample, we've also decided to be wrong 5% of the time, and we know we want to estimate a range of values. Therefore, we know we want to use a confidence interval with an alpha of 0.05. If we have interval-ratio data, and have a sample mean, we use the following equation:

---

[1] Blalock, Hubert M, Jr. 1979. Social Statistics. New York: McGraw-Hill, Inc.

$$CI=\overline{X}\pm Z(\sigma/\sqrt{N})$$

This equation estimates the population interval, given the sample mean and the standard deviation of the sampling distribution (sigma). Since we rarely know the latter, we can estimate sigma with the standard deviation of the sample (s), and use the following equation:

$$CI=\overline{X}\pm Z(s/\sqrt{N-1})$$

This equation allows us to estimate the population interval with the sample mean and the sample standard deviation. What we find, then, is a range of values in which we are 95% confident that the population mean will fall between. Also, though, there is a 5% chance that the population mean does <u>not</u> fall within this range.

We can also estimate the population value for nominal data by using proportions. Let's say 9% of our sample of 250 respondents are black and we want to estimate the proportion of blacks in the population. Using the following equation, we can get this estimate:

$$CI=P_s\pm Z\sqrt{\frac{P_u(1-P_u)}{N}}$$

Our sample proportion (P_s) is 0.09, with an alpha of 0.05, Z is 1.96, and we set the population proportion (P_u) to be 0.05. We can then solve this equation:

$$CI=0.09\pm 1.96\sqrt{\frac{0.5(1-0.5)}{250}}=0.09\pm 0.06$$

Therefore, we can infer from our sample that the proportion of blacks in the population is between 0.03 and 0.15.

Remember that the width of the intervals is dependent on two things: the alpha level and the sample size. The more errors you are willing to make (the higher the alpha level), the smaller your interval width. Thus if you are willing to be wrong 10% of the time, your interval width will be smaller than if you are willing to be wrong only 1% of the time.

Also, increased sample size will decrease your interval width. As your sample size increases, you are collecting a wider range of data on the population, and thus you are able to make a better prediction of the population value.

MULTIPLE-CHOICE QUESTIONS

1. A point estimate is:
   a. an estimate of the sample distribution
   b. an estimate of the sample size
   c. an estimate of the population size
   d. an estimate of the population value

2. A confidence interval is:
   a. an estimate of the sample distribution
   b. an estimate of the sample size
   c. an estimate of the population size
   d. an estimate of the population value

3. The main difference between a point estimate and a confidence interval is that point estimates use _________ whereas confidence intervals use _________.
   a. a point, a range
   b. an estimate, an alpha level
   c. a range, a point
   d. a population, a sample

4. A good estimator should be ___________ and __________.
   a. biased, efficient
   b. biased, evasive
   c. unbiased, inefficient
   d. unbiased, efficient

5. A statistic is unbiased if:
   a. the mean of the sampling distribution is equal to the sample
   b. the sample size is larger than 50
   c. the mean of the sample is equal to the mean of the population
   d. the mean of the sampling distribution is equal to the population

6. Efficiency of a measure indicates:
   a. the degree to which the population mean is equal to the standard deviation of the sample means
   b. the degree to which the sample outcomes are clustered around the mean of the sampling distribution
   c. the distance between the population value and the sample value
   d. the population value can be found quickly and easily

7. Alpha levels are:
   a. the amount of error in a sample
   b. the probability of the population
   c. the probability of error
   d. the Greek letter for population variance

8. Common alpha levels are:
   a. .50, .25, .15
   b. .10, .01, .15
   c. .99, .95, .90
   d. .01, .05, .10

9. To find the confidence interval for population means, we must have:
   a. sample mean and standard deviation
   b. population mean and standard deviation
   c. sample mean and proportion
   d. population mean and sample standard deviation

10. To find the confidence interval for population proportions, we must have:
   a. sample proportion and standard deviation
   b. sample mean and standard deviation
   c. sample proportion and sample size
   d. population mean and sample standard deviation

11. The width of confidence intervals is affected by _________ and __________.
   a. sample size, sample means
   b. alpha levels, sample proportions
   c. sample means, standard deviations
   d. sample size, alpha levels

12. The larger the sample size, the __________ the confidence interval.
   a. smaller
   b. wider
   c. more continuous
   d. greater

13. The lower the confidence interval, the ________ the interval width.
   a. smaller
   b. wider
   c. more continuous
   d. more precise

14. A confidence level of 0.01 has a Z score of:
   a. $\pm$ 3.25
   b. $\pm$ 2.58
   c. $\pm$ 1.96
   d. $\pm$ 1.65

15. A confidence level of 0.05 has a Z score of:
   a. $\pm$ 3.25
   b. $\pm$ 2.58
   c. $\pm$ 1.96
   d. $\pm$ 1.65

WORK PROBLEMS

1. Construct a confidence interval for each of the sample outcomes below at the 0.01 and 0.05 levels. Discuss the changes for each.

a. $\bar{X}$=34.6
   s=12.3
   N=187

b. $\bar{X}$=68
   s=25
   N=540

c. $\bar{X}$=164.8
   s= 83.2
   N=150

d. $\bar{X}$=76
   s= 9.6
   N=100

e. P_s=0.35
   N =112

f. P_s=0.56
   N =456

2. Find the corresponding Z score for each confidence level below:

Confidence Level	Alpha	Z	Area Beyond Z Score
90%	0.10	0.0505	1.64
95%			
99%			
80%			
75%			

3. In a study of attitudes toward smoking, a researcher found that 75 of his 238 respondents smoke. Estimate the number of smokers in the population at an alpha of 0.05.

4. In a sample of all majors at a large university (N=1200), the grade point average was 3.2, with a standard deviation of 0.9. Estimate the grade point average for the entire university, at a 90% confidence level.

5. The average grade of a sample of 125 students in a statistics course at a university was 78.5, with a standard deviation of 15.7. What is the population average for an alpha of 0.05?

6. Out of a sample of 325, 175 respondents stated that they were in favor of women's rights. What proportion of the population favors women's rights, at the 95% confidence interval?

7. A researcher found in a sample of 238 households that the average income is $31,457, with a standard deviation of $15,722. Estimate the population income at a 99% confidence level.

8. In a sample of 125 inmates at a local prison, the average sentence was 8 years, with a standard deviation of 3 years. Estimate the average sentence for the population of all prison inmates at the 0.05 alpha level.

9. In a similar study the same mean and standard deviation were found, but the sample size was 1200. How does this change the results. Explain.

10. In a sample of 450 graduating seniors at Lowdown College, 289 have found jobs. What proportion of the population have jobs?

11. In a study of women's health, it was found that on average, women went to see a physician for a regular check up 1.3 times per year (s=4.3). From this sample of 300 women, estimate the number of physician visits for the entire population, with only a 1% chance of error.

12. A survey of the men's basketball teams at some large universities found that the average height was 6'2" (74 inches), plus or minus 6 inches. With a sample size of 135, estimate the height of all male basketball players.

13. In a survey of attitude toward racially mixed marriages, a researcher found that the majority of her sample (235 out of 300) did not have a problem with them. Given a 90% confidence level, what proportion of the population have the same attitude?

14. You are asked by the administration of your college to use your new knowledge of statistics to conduct research on the campus regarding sexuality. The university is primarily concerned with unwanted pregnancy rates and wants to be very certain of the results (95% confident) with desired interval width of $\pm$ 2%. How large a sample size would you recommend?

15. A drug company is testing their new allergy medicine. Of the 1750 people being tested, 123 had negative reactions. Since they are very concerned about the health and well-being of their customers, they want to be very sure of their results. At a 99% confidence level, estimate the proportion of negative reactions in the population.

DETAILED ANSWERS

1.
a. 34.6 $\pm$ 1.76
b. 68 $\pm$ 2.11
c. 164.8 $\pm$ 13.36
d. 76 $\pm$ 1.90
e. 0.35 $\pm$ 0.098
f. 0.56 $\pm$ 0.04

2.

Confidence Level	Alpha	Z	Area Beyond Z Score
90%	0.10	0.0505	1.64
95%	0.05	0.0250	1.96
99%	0.01	0.0049	2.58
80%	0.20	0.0985	1.29
75%	0.25	0.1230	1.16

3. Because this is nominal data, you want to find the proportion of smokers in the sample which is 75/238, or 0.32. The estimate of the proportion of smokers in the population then is 0.32 $\pm$ 0.06. Said another way: The total proportion of smokers in the population is estimated to be between 0.26 and 0.38.

4. Since GPA is an interval-ratio variable, we want to estimate the population mean, which is 3.2 $\pm$ 0.05, or between 3.15 and 3.25. Notice how small the interval width is. This is a product of the large sample size.

5. Therefore, the average grade in the population is estimated to be between 75.7 and 81.26.

$$CI = \overline{X} \pm Z(s/\sqrt{N-1}) = 78.5 \pm 1.96(15.7/\sqrt{124}) = 78.5 \pm 2.76$$

6. The proportion of respondents who are in favor of women's rights is 175/325, or 0.54. To find the population proportion:

$$CI = P_s \pm Z\sqrt{\frac{P_u(1-P_u)}{N}} = 0.54 \pm 1.96\sqrt{\frac{0.5(1-0.5)}{325}} = 0.54 \pm 0.06$$

Therefore, between 48% and 60% of the population are in favor of women's rights.

7.

$$CI = \overline{X} \pm Z(s/\sqrt{N-1}) = 31457 \pm 2.58(15722/\sqrt{237}) = 31457 \pm 2635.6$$

The average income in the populations is between $28,821.40 and $34,092.60.

8.

$$CI = \overline{X} \pm Z(s/\sqrt{N-1}) = 8 \pm 1.96(3/\sqrt{124}) = 8 \pm 0.53$$

Therefore, the average sentence for the population is between 7.47 years and 8.53 years.

9.

$$CI = \overline{X} \pm Z(s/\sqrt{N-1}) = 8 \pm 1.96(3/\sqrt{1199}) = 8 \pm 0.09$$

With the increased sample size, we are better able to predict the value of the population, therefore the width of the interval is reduced. The average sentence for the population is between 7.82 years and 8.18 years.

10. The proportion of graduating seniors who have found employment is 289/450, or 0.64. To find the population proportion:

$$CI = P_s \pm Z\sqrt{\frac{P_u(1-P_u)}{N}} = 0.64 \pm 1.96\sqrt{\frac{0.5(1-0.5)}{450}} = 0.64 \pm 0.05$$

Therefore, between 59% and 69% of the population of graduating seniors have found employment.

11.

$$CI = \overline{X} \pm Z(s/\sqrt{N-1}) = 1.3 \pm 2.58(4.3/\sqrt{299}) = 1.3 \pm 0.65$$

The population of women see a physician between 0.65 and 1.95 times per year.

12.

$$CI = \overline{X} \pm Z(s/\sqrt{N-1}) = 74 \pm 1.96(6/\sqrt{134}) = 74 \pm 1.02$$

The average height for all basketball players then is between 6'1" and 6'3".

13.

$$CI = P_s \pm Z\sqrt{\frac{P_u(1-P_u)}{N}} = 0.78 \pm 1.65\sqrt{\frac{0.5(1-0.5)}{300}} = 0.78 \pm 0.05$$

Therefore between 73% and 83% of the population do not have a problem with racially mixed couples.

14. If:

$$CI = P_s \pm Z\sqrt{\frac{P_u(1-P_u)}{N}}$$

Then:

$$0.02 = Z\sqrt{\frac{P_u(1-P_u)}{N}}$$

Therefore, we would need a sample size of approximately 2400 in order to have an interval width of ± 0.02.

15.

$$N = \frac{Z^2[P_u(1-P_u)]}{0.02^2} = \frac{3.84(0.05)}{0.0004} = 2401$$

$$CI = P_s \pm Z\sqrt{\frac{P_u(1-P_u)}{N}} = 0.07 \pm 2.58\sqrt{\frac{0.5(1-0.5)}{1750}} = 0.07 \pm 0.03$$

Approximately between 4% and 10% of the population is estimated to have a negative reaction to the medication.

CHAPTER EIGHT
HYPOTHESIS TESTING I:
The One-Sample Case

In chapter eight of your textbook, you are introduced to the idea of hypothesis testing. In this particular type of hypothesis testing, we are trying to discern whether or not our sample is really from the population, or from some other population. Sounds strange, doesn't it? But, follow along, and you'll probably agree that this is an important concept!

Basically it works like this: let's say we are interested in height and shoe size, but we think that basketball players will be much taller and have much larger feet than the rest of the population. In fact, you could consider all basketball players as being a separate population from all other people, non-basketball players. Then we could take a sample from the population of all basketball players, and compare the average height of this sample of the population to the average height of the other population (non-basketball players).

Therefore, you are comparing a sample mean, $\bar{x}$ (average height of basketball players) to a population mean, μ (average height of the population of non-basketball players), in order to find out whether or not the population of basketball players is different from the population of non-basketball players. If we find a **significant** difference, then we decide that our sample of basketball players really are from a different population than the rest of the population.

Let me stress again: we are trying to decide whether or not our sample of basketball players is from the same population as non-basketball players, or if our sample is so different that we have to decide that when looking at height, the population of basketball players should be considered different from the non-basketball player population.

I know it sounds a little crazy when you first hear about it, but hypothesis testing for one-sample means and proportions is very important. Consider when a policy maker is trying to decide who gets what money. If you can say to him/her that Hispanics are getting significantly less money for education than non-Hispanics, than he/she knows that this population (Hispanics) needs more money than the other population (non-Hispanics), and furthermore that Hispanics should be considered as a different population on this variable than non-Hispanics. In other words, Hispanics do not belong to the same population as non-Hispanics, and should be considered different on this variable.

In order to test one-sample hypotheses, you need to follow the five-step model for hypothesis testing. These five steps allow you to make some assumptions about your data, sampling distribution, and your sample, as well as to define your null and research hypotheses, and to determine your critical region.

When stating your hypotheses, you need also to decide whether you think your sample is merely different from the population, or do you think perhaps the sample is higher or lower than the population on the variable in question. With this decision, you set your test statistic to be either one-tailed or two-tailed. In one-tailed tests, you are hypothesizing the direction of the difference: basketball players will be **taller** than non-basketball players.

In two-tailed tests, you are not hypothesizing the direction of the difference, only that the difference will exist: basketball players will be different from the population of non-basketball players, but I am unsure whether they will be taller or shorter.

If the basketball players' average height is higher than that of non-basketball players, the difference is on the positive side of the mean. If the basketball players' average height is lower, then the difference is on the negative side of the mean.

Remember: the differences in the populations is compared with Z scores. In Z scores, we are transforming the value of the scores so that the mean is zero and the standard deviation is one. In hypothesis testing, the population mean (μ) is zero, and we compare our sample mean($\bar{x}$) to that point. In order to do so, the following equation is used:

$$Z=\frac{\overline{X}-\mu_{\overline{x}}}{\sigma/\sqrt{N}}$$

In this equation, the mean of the sample is compared to the mean of the sampling distribution (which is equal to the population mean), and divided by the standard deviation of the sampling distribution. This equation then is finding the Z score for the sample mean so that we can compare it with the population mean to determine if there is a significant difference between the two.

Many times, though, the standard deviation of the population (σ) is not known; therefore, we need to estimate it with the sample standard deviation (s). If the sample size is large (larger than 100) this can be done by substituting s for σ, and using the following equation:

$$Z=\frac{\overline{X}-\mu_{\overline{x}}}{s/\sqrt{N-1}}$$

Remember: when estimating the standard deviation of the population (σ) with the standard deviation of the sample (s), we need to correct for the measure by using N-1 rather than just N.

If the sample size is small, though, we need to use what is called **Student's t distribution**. Before we get into the test statistic, how about a fun little story about "Student." My students always ask: Why is it called "Student's t?" and "Who is Student?" As it happens, "Student" was a scientist named William Gossett at the Guinness Brewery in Dublin, Ireland at the turn of the century who was in charge of testing the yeast culture in the beer. Because he was using very small sample sizes to estimate the yeast, he discovered that he was making many errors in his estimations. Therefore, he created this test statistic in order to control for the smaller sample size. Since he was still employed by Guinness Brewery, he couldn't publish this new statistic under his name without letting the competing breweries know what he was up to, so he called it "Student's t" (Schutte, 1973[1]). A nice piece of trivia for you. All right, back to Student's t. The t-distribution takes into account small sample sizes, since the smaller the sample size, the less representative it is of the population. For example, if you ask 20 people how much money they make, the average of these 20 incomes is not as accurate a

---

[1] Schutte, Jerald G. 1977. Everything You Always Wanted to Know about Elementary Statistics (but were afraid to ask). Englewood Cliffs: Prentice-Hall, Inc.

measure of income as if you had asked 200 people. Therefore, the distribution of the sample will not be normal in shape, but instead flatter. Therefore, Student's t-distribution takes into account this flatter shape, and bases the test statistic on this distribution. The test statistic for Student's t is:

$$t(obtained) = \frac{\bar{X} - \mu}{s/\sqrt{N-1}}$$

This equation is basically the same as Z (obtained) when estimating the standard deviation of the sampling distribution (σ). The main difference between Z and t is the critical region, which is smaller for the t distribution than it is for the Z distribution. This again is a product of small sample sizes: we are not convinced that a sample of less than 100 is a true representation of the population, therefore, t (critical) corrects for this problem.

Another important point to consider whenever you are using hypothesis tests is the possibility of error. We know that if we set the alpha level to be 0.05, that we are expecting to make errors 5% of the time. More precisely, when we reject a null hypothesis or fail to reject a null hypothesis, we know that we are only 95% confident about these conclusions; therefore there is the possibility (albeit a small one) that we are wrong.

These possibilities are called Type I and Type II errors, or alpha and beta errors, respectively. **Type I**, or **alpha errors**, are made when we reject a null hypothesis which is really true. Or, in English: we conclude that there really is a significant difference between our sample and the population when in fact there isn't. Remember: we are only as certain as our alpha level; so we know there's a 5% chance of being wrong--this is that 5%.

Type II, or **beta errors**, are made when we fail to reject a null hypothesis that is really false. In English this means that we decide that there is no significant difference between our sample and the population, when in actuality there is.

Think of it this way: There is only one correct answer in the entire world -- either there is a significant difference between our sample and the population or there is not. We are 95% sure that our results are true, but there is a 5% chance that we have made **either** an alpha error **or** a beta error.

The problem with both alpha and beta errors (besides being sticky little buggers) is that you really never know when you are committing one, only that there is a 5% chance that we have made one. Also, if you decrease the likelihood of committing one type of error, you are increasing the likelihood of committing the other type of error. As a general rule, you want to minimize the type one (alpha) error, rather than the type II (beta) error. Thus, we set our alpha levels to be low, reducing the likelihood of rejecting a null hypothesis that is actually true.

MULTIPLE-CHOICE QUESTIONS

1. The main point in one-sample hypothesis testing is to:
   a. determine if there is more than one sample
   b. determine if the sample is taken from the population
   c. determine if one sample is large enough to estimate the population value
   d. determine the sample size necessary to estimate the population

2. The Z and t distributions are:
   a. exactly the same
   b. different, because Z is flatter than t
   c. different because t is flatter than Z
   d. incomprehensible

3. Z (critical) is:
   a. the test statistic computed in step four of the hypothesis process
   b. the score that marks the mean
   c. the test that determines the sampling distribution
   d. the score that marks the beginning of the critical region

4. Z (obtained) is:
   a. the test statistic computed in step four of the hypothesis process
   b. the score that marks the mean
   c. the test that determines the sampling distribution
   d. the score that marks the beginning of the critical region

5. The critical region is:
   a. the area between the mean and the obtained value
   b. the area between the mean and the critical value
   c. the area beyond the critical value
   d. the area given in Appendix B

6. If you hypothesized that the sample mean was greater than the population mean, you would need to use ____________________________.
   a. one-sample proportion test
   b. two-tailed test
   c. one-tailed test
   d. population estimators

7. The statement in which you say there is "no difference" between the sample and the population is called the ________________.
   a. null hypothesis
   b. research hypothesis
   c. assumptions statement
   d. critical region

8. The statement in which you say there is a difference between the sample and the population is called the ________________.
   a. null hypothesis
   b. research hypothesis
   c. assumptions statement
   d. critical region

9. If you state that the sample mean is hypothesized to be less than the population mean, then the critical region should be:
   a. on both sides of the mean
   b. on the right side of the mean
   c. on the left side of the mean
   d. in the center of the distribution

10. If you have a sample size larger than 100 and a known standard deviation of the population (σ), you would use ______________ to test for significance.
    a. Z distribution for one-sample proportions
    b. t distribution for one-sample proportions
    c. Z distribution for one-sample means
    d. t distribution for one-sample means

11. If you have a sample size smaller than 100, and an unknown standard deviation of the population (σ), you would use ______________ to test for significance.
    a. Z distribution for one-sample proportions
    b. t distribution for one-sample proportions
    c. Z distribution for one-sample means
    d. t distribution for one-sample means

12. If you want to test for significance for nominal data, you would use______________.
    a. Z distribution for one-sample proportions
    b. t distribution for one-sample proportions
    c. Z distribution for one-sample means
    d. t distribution for one-sample means

13. When testing for significance between a sample and a population in nominal data, the term P_u is the ________________.
   a. sample proportion
   b. population proportion
   c. sample mean
   d. population mean

14. The probability of failing to reject a null hypothesis that is, in fact, false, is a ______________.
   a. type I, or alpha error
   b. type II, or beta error
   c. critical value
   d. critical region

15. The probability of rejecting a null hypothesis that is, in fact, true is a ______________.
   a. type I, or alpha error
   b. type II, or beta error
   c. critical value
   d. critical region

WORK PROBLEMS

Solve each of these problems, using the five-step hypothesis model whenever appropriate. Unless otherwise specified, use an alpha level of 0.05.

1. A researcher was interested in the GPAs of athletes. He found that in his sample of 350 varsity athletes, the average GPA was 2.9. He knows the overall average GPA is 3.1, with a standard deviation of 2.3. Is there a significant difference between athletes and the rest of the population?

2. The average years of education for a sample of 50 CEOs of the top corporations is 17.5 years, with a standard deviation of 1.2 years. The overall average years of education is 13.4 years. Is the difference significant?

3. The proportion of Hispanic students in a survey of 143 students at a college campus in south Texas was 0.35. The proportion of Hispanic students in all universities is 0.09. Does this college have significantly more hispanics than the rest of the college campuses?

4. The average income in a rural community was found to be $30,900 (s=$5,940; N=250). The average income in the population is $30,765. Is this community significantly different from the rest of the population?

5. In a study of attitudes toward women working outside the home, a researcher found that 357 men were in favor, while 298 men opposed. In the population, the proportion of people who favor women working outside the home is 0.47. Can we say that the men are significantly different?

6. In a recent drug test for a new allergy medicine, the researchers found that of those who were given the medicine (1500 patients), only 6.2% of them reported drowsiness. The drowsiness rate for all other allergy medicines is 10.5%. Because this research is concerned with medication, the researchers wanted to be 99% confident of their findings. Is this medicine significantly different?

7. A recent study found that there was no significant difference between a sample of prostitutes and the rest of the population for unwanted pregnancies. If there were 27 prostitutes in the sample, what would be the critical value which would be needed for significance to be found? [Hint: the researchers hypothesized that prostitutes would have a higher rate of unwanted pregnancies than the rest of the population].

8. In a study of 57 homes surrounding a community park, a researcher found that, as he hypothesized, the residents had a lower tolerance for noise than the rest of the city. The t-(obtained) for the difference was -2.55. What can you conclude? How would his findings change if he set the alpha level to be 0.01? How about 0.10?

9. A group of researchers were interested in modern contraceptive use in a rural village in the mountains of Bolivia, and found that of their sample of 174 women, 34 of the women had knowledge about modern contraceptives. This number was smaller than the estimated 70% of the rest of Bolivian women who had knowledge of modern contraceptives. Is the village from the same population?

10. In a study of farming practices, it was found that those farmers who used hybrid corn seed had an average yield of 50 bushels (N=250). All other farmers had an average yield of 47 bushels (with a standard deviation of 10.86 bushels). Given an alpha level of 0.01, is there a significant difference between the hybrid seed farmers and the rest of the population?

11. A researcher hypothesized that people with AIDS would have a lower quality of life than people without AIDS. She collected data using a Quality of Life scale and found that the average score of the 216 AIDS patients was 6.7, with a standard deviation of 4.3. She knew that the rest of the population had an average score of 7.1. Do AIDS patients have significantly lower quality of life than the rest of the population? [Assume Quality of Life to be interval-ratio]

12. The average number of friends a sample of 47 college women reported as having was 4.9, with a standard deviation of 1.4. The national average of friends women report having is 3.2. Do women in college report having significantly more friends than the rest of the population?

13. The number of dates the men's dorm of Wesaloosers was pathetically low. Boredom got the best of Tom one long, lonely Friday night, so he decided to take a poll of 30 of his fellow dorm-mates regarding the number of dates each had had in the last month. The average number of dates was 2.1, plus or minus 0.4 dates. He knew from a report recently published by the university that the average number of dates for all other dorm-dwellers was 4.6 per month. Are the men of Wesaloosers significantly losing out on dates?

14. Lonely Tom, from Question 13, consoled himself, and other Wesaloosers dorm-mates, by comparing their GPAs with the overall campus GPA. The Wesaloosers dorm sample of the same 30 men had an average GPA of 3.6, plus or minus 0.7. The GPA for the entire campus was 3.1. Can poor Tom take solace in the fact that their dorm grades are higher?

15. Tom still can't figure out why his dorm has the fewest dates, so he then figures out that his dorm of 340 men spend an average of 4 nights per week in the library. The campus average is 2.5 nights per week. Are they spending <u>too</u> much time in the library?

DETAILED ANSWERS

1. The average GPA for the athletes is 2.9, while for the rest of the population is 3.1. Solving for Z, the equation would be:

$$Z=\frac{2.9-3.1}{2.3/\sqrt{350}}=-1.63$$

Z(critical) is $\pm$1.96. Since -1.63 does <u>not</u> fall in the critical region, there is no significant difference between the athletes and non-athletes with regard to GPA.

2. Because of the small sample size, N=50, the t distribution should be used. Solving for t:

$$t=\frac{17.5-13.4}{1.2/\sqrt{49}}=24.12$$

In order to find t(critical) a two-tailed test, at the 0.05 level, you need to find the degrees of freedom (df), which is equal to N-1, or 50-1=49. If you look down the df column, you will see that the numbers skip from 40 to 60. In cases such as these, you want to choose the **larger** of the critical values. For df=40, t(critical) is 2.021; for df=60, t(critical) is 2.000. Therefore, you should go with the larger critical value of 2.021.

Regardless, though, t(obtained), 24.12, falls inside the critical region. Therefore, there is a significant difference between the education of CEOs and the rest of the population.

3. The sample proportion (P_s) = 0.35; the population proportion (P_u) = 0.09; using the Z equation for sample proportions:

$$Z = \frac{P_s - P_u}{\sqrt{\frac{P_u(1-P_u)}{N}}}$$

Solving for the unknown, we get the following equation:

$$Z = \frac{0.35-0.09}{\sqrt{\frac{0.09(1-0.09)}{143}}} = \frac{0.26}{0.02} = 13$$

Z(obtained), then, is 13; because we hypothesized that this college had significantly more Hispanic students than the rest of the colleges, we are interested in only the positive scores. Thus, we have a one-tailed test, which gives us a Z(critical) of +1.65.

Our Z(obtained) falls in the critical region, therefore, we can say that the population at this college has significantly more Hispanic students than other colleges.

4. Solving for the unknown Z value, we would use the following equation:

$$Z = \frac{30,900-32,765}{5,940/\sqrt{250}} = -4.95$$

Since we hypothesized that there would be a difference, not a direction to this difference, we will consider the areas in both sides of the tail. The Z (critical) for a two-tailed test is $\pm$ 1.96; -4.95 falls in the critical region, therefore there is a significant difference between this community and the rest of the population.

5. If 357 men are in favor and 298 are opposed, we have a sample size of 655 (357+298). We are interested in the proportion of those in favor, which is 0.55 (357/655 = 0.55). The overall proportion is 0.47. To solve for Z, use the following:

$$Z=\frac{0.55-0.47}{\sqrt{\frac{0.47(1-0.47)}{655}}}=4.0$$

Being a two-tailed Z test, with an alpha level 0.05, Z(critical) is ±1.96. Z(obtained) does fall in the critical region, therefore we reject the null hypothesis of no difference. Instead, men are significantly different than the rest of the population concerning attitudes toward women working outside the home.

6. In this medical study, 6.2% of the patients reported drowsiness with this new medicine; that is a proportion of 0.062, compared to 0.105 of the patients using all other allergy medicines. Solving for Z, the Z(obtained) is -5.44, which is in the critical region for a two-tailed test at the 0.01 alpha level, which is ±2.58.

7. With a sample of only 27 prostitutes, we need to use the t distribution. The degrees of freedom is N-1, or 26. At an alpha of 0.05, given a one-tailed test, the t(critical) is +1.706. Therefore, the t(obtained) would have to have been larger than +1.706 to be significant.

8. Given a one-tailed test, at the 0.05 level, with a degrees of freedom of N-1, or 56, the critical value is between 1.684 and 1.671. Since we want to use the largest critical value to decrease the likelihood of type I alpha error, we set our t(critical) to be 1.684. Since it was hypothesized that the park residents were less tolerant, the critical region will be in the lower tail, and the critical value will be -1.684.

Therefore, we can conclude that the park residents do have significantly lower tolerance for noise than other residents in the city.

9. If 34 out of 174 women had knowledge of modern contraceptives, that is approximately 20%, or a proportion of 0.20. Comparing that to the 70% of women in the rest of the population yields a Z(obtained) of +16.67. This value falls into the critical region, therefore, there is a significant difference between the women in this village and the rest of the population.

10. Z(critical) for an alpha of 0.01, two-tailed test is ±2.58. The Z(obtained) is +4.35, and falls in the critical region. Therefore, we reject the null hypothesis: there is a significant difference in crop yields. Farmers who used the hybrid corn seed have significantly higher crop yields than farmers who do not.

11. It is hypothesized that the AIDS patients will have significantly lower quality of life than the rest of the population. Therefore we are interested in the lower tail of the curve only, at an alpha level of 0.05. The critical value then is -1.65. The obtained value is -1.38. Therefore, this value does not fall in the critical region. There is no significant difference between the AIDS patients and the rest of the population on the variable quality of life.

12. Yes, college women report significantly more friends than the rest of the population. T(critical) for this sample is +1.684; t(obtained) is +8.10.

13. Poor Wesaloosers guys. Yes, they have significantly fewer dates than the rest of the university students. Because of the small sample size, the t distribution should be used. The t(critical) is -2.045, the t(obtained) is a whopping -35.71.

14. The Wesaloosers can take heart in the fact that their GPAs are significantly higher than the rest of the university. The critical value is +2.045, and the obtained value is +3.85. Therefore, their GPAs are significantly higher.

15. If the Wesaloosers dorm averages 4 out of 7 nights in the library, that gives them a proportion of 0.57. The rest of the university students spend an average of 2.5 nights per week in the library (0.36). Using the Z equation for one-sample proportions, Z(obtained) is +5.25, and Z(critical) is +1.65. Yes, the Wesaloosers do spend more time in the library than all other students. Maybe they should get out more!

CHAPTER NINE
HYPOTHESIS TESTING II:
The Two-Sample Case

The logic behind two-sample hypothesis testing is very similar to the one-sample tests, but instead of testing for difference between your sample and the rest of the population, you are testing for difference between two populations. Specifically, you are using two **independent random samples** of two different populations (ie. male, female; high income households, low income households). So although you are using two samples, you are estimating **two populations**.

This is a very important distinction. Remember: we are using **samples** of the two populations, but making our hypotheses about the **populations**. I know this sounds funny and strange, but try to think of both groups, whether they be males and females, or Protestants and non-Protestants, as **two distinct populations**, which we are hypothesizing (in the research hypothesis statement) to be different from one another on some variable.

There are three[1] types of two-sample hypothesis tests: two-sample means for large samples (Z-distribution); two-sample means for small samples (t-distribution); and two-sample proportions for large samples (Z-distribution). The main distinction in all of these is whether you have means (interval-ratio data) or proportions (nominal data), and are your samples large (combined total greater than 100) or small (combined total less than 100). If you know the answers to each of these, you can easily find the correct test to use.

In the null hypothesis statement for two-sample means, you are stating that there is no difference between the two populations, which is stated as:

$$H_o: \mu_1 = \mu_2$$

In the null hypothesis statement for one sample means, you are simply stating that there is no difference between your **sample** and the rest of the population. Similarly, in the research hypothesis, you are stating either that the two populations are different from one another:

$$(H_1: \mu_1 \neq \mu_2)$$

or that one population is either greater than or less than the other population:

$$(H_1: \mu_1 > \mu_2)$$

---

[1] Your textbook also mentions a fourth hypothesis test, matched samples, but it is an optional section, so it will not be discussed here.

$$(H_1: \mu_1 < \mu_2)$$

Note that in all of these statements we are stating that the populations are either similar or different, not the samples!

MULTIPLE-CHOICE QUESTIONS

1. In the two-sample case of hypothesis testing, we are comparing a ____________ to a ____________.
   a. sample, sample
   b. sample, population
   c. population, sample
   d. population, population

2. In order to test for significance between the two populations, we assume:
   a. simple random samples
   b. independent random samples
   c. sampling distribution to be random
   d. sampling distribution to be positive

3. One main differences between one-sample hypothesis testing and two-sample hypothesis testing is that in the latter we are:
   a. comparing means
   b. comparing samples
   c. comparing a sample to the population
   d. comparing separate populations

4. In the assumption statement for two-sample means (small samples), we state:
   a. independent random samples
   b. level of measurement is interval-ratio
   c. equal population variances
   d. all of the above

5. In the assumption statement for two-sample proportions, we state:
   a. random sampling
   b. level of measurement is normal
   c. sampling distribution is normal
   d. all of the above

6. The null hypothesis states:
   a. no difference between samples
   b. no difference between populations
   c. the two samples are different
   d. the two populations are different

7. The research hypothesis states:
   a. no difference between samples
   b. no difference between populations
   c. the two samples are different
   d. the two populations are different

8. The Z-distribution is used for ______ samples, whereas the t-distribution is used for ________ samples.
   a. nominal, interval-ratio
   b. interval-ratio, nominal
   c. small, large
   d. large, small

9. The equation for degrees of freedom for two-samples is:
   a. N -2
   b. $N_1 (N_1 - 1)$
   c. $N_1 - N_2$
   d. $N_1 + N_2 - 2$

10. When calculating the Z score for two-sample proportions, P_u is:
   a. calculated first
   b. given
   c. never used
   d. all of the above

11. In the equation for two-sample means, the term $(\mu_1 - \mu_2)$ is equal to ________.
   a. the sample means
   b. zero
   c. not in this equation
   d. used only in small samples

12. If a value was found to be significant, we would conclude:
   a. we fail to reject the null hypothesis
   b. we reject the null hypothesis
   c. there is no difference between the two populations
   d. there is a significant difference between the two samples

13. If we had hypothesized a one-tail test for a large, two-sample means test at an alpha of 0.05, and the Z(obtained) calculated to be +1.62, we would conclude:
a. we fail to reject the null hypothesis
b. we reject the null hypothesis
c. there is a significant difference between the two populations
d. there is a significant difference between the two samples

14. For a two-sample proportions test, we need:
a. sample sizes less than 100
b. the standard deviation of the sampling distribution
c. degrees of freedom
d. large sample sizes

15. In order to test the difference between males and females with regard to whether they favor or oppose the death penalty, we would use:
a. two-sample means, large samples
b. two-sample means, small samples
c. two-sample proportions, large samples
d. two-sample proportions, small samples

WORK PROBLEMS

1. Is there a significant difference in incomes between men and women at the 0.05 level?

Income

Men		Women	
$\bar{X}_1=$	\$30,450	$\bar{X}_2=$	\$29,950
$s_1=$	1,595	$s_2=$	1,790
$N_1=$	32	$N_2=$	27

2. The following data are for the number of religious services attended per year for Protestants and Catholics. Do Catholics attend more services than Protestants?

Number of Services Attended

Protestants		Catholics	
$\bar{X}_1=$	9.5	$\bar{X}_2=$	11.3
$s_1=$	3.1	$s_2=$	4.7
$N_1=$	147	$N_2=$	176

3. A Sociology professor believes that Sociology majors have higher GPAs than Psychology majors. Given the data below, is the difference significant?

GPA			
Sociology Majors		Psychology Majors	
$\bar{X}_1=$	3.1	$\bar{X}_2=$	3.0
$s_1=$	1.4	$s_2=$	1.3
$N_1=$	250	$N_2=$	300

4. Educational achievement was measured for both lower income and higher income individuals. Is there a significant difference?

Educational Achievement			
Lower Income		Higher Income	
$\bar{X}_1=$	12.5	$\bar{X}_2=$	14.3
$s_1=$	3.4	$s_2=$	2.5
$N_1=$	38	$N_2=$	47

5. In a survey of attitudes, 42% of the suburban respondents (N=142) said they were afraid of crime. Compare that to the 89% (N=212) of the city respondents. Are city-dwellers more afraid of crime?

6. What is the critical value for two samples ($N_1=15$, $N_2=21$) for a two-tailed test, with an alpha level of 0.01?

7. A researcher found a t(obtained) for a two-sample means test of 2.798. She was interested in a significance level of 0.01 for the one-tailed test. The sample sizes were: $N_1=11$, $N_2=12$. Is the test statistic significant?

8. In a study of women's knowledge of modern contraceptives in South America, a researcher has samples from Bolivia and Peru. The data below are the proportion of women who knew about modern contraceptive methods. Is there a significant difference?

Knowledge of Contraceptives	
Bolivia	Peru
$P_1=0.71$	$P_2=0.69$
$N_1=1350$	$N_2=1205$

9. In a study of quality of life, a researcher found a difference in scores by racial groups. Are these differences significant at the 0.01 level?

Quality of Life

White		Black	
$\bar{X}_1=$	6.9	$\bar{X}_2=$	7.2
$s_1=$	1.4	$s_2=$	1.7
$N_1=$	132	$N_2=$	42

10. Of the recently graduating class, the average income of starting salaries is given below. Do Physical Science majors make significantly more than Social Science majors?

Salaries

Physical Science		Social Science	
$\bar{X}_1=$	\$30,472	$\bar{X}_2=$	\$28,800
$s_1=$	3,273	$s_2=$	2,960
$N_1=$	785	$N_2=$	750

11. The administration at a large university was worried about the Greek system: specifically, do members of sororities and fraternities spend less hours studying than non-members? A survey of both groups revealed the following data.

Hours of Study

Greek		Non-Greek	
$\bar{X}_1=$	12.3	$\bar{X}_2=$	14.0
$s_1=$	4.1	$s_2=$	0.9
$N_1=$	51	$N_2=$	53

12. A couple is planning to move to a new city several hundred miles away. They want to move into an area with lots of social activities so they can meet friends quickly. They have looked at several neighborhoods in both the suburbs and the city and have collected data on number of neighborhood parties in each. Which would you suggest they move to?

Number of Parties

Suburbs		City	
$\bar{X}_1=$	4.2	$\bar{X}_2=$	3.5
$s_1=$	1.3	$s_2=$	0.9
$N_1=$	14	$N_2=$	12

13. The guys in Wesaloosa, whom we met in chapter eight, are still upset by their dating record, but even more so now. The guys in Usaloosa dorm have done a little survey of their own dating habits. The following data were collected. Is there a significant difference in the number of dates?

Number of Dates

Wesaloosa Dorm		Usaloosa Dorm	
$\bar{X}_1=$	2.1	$\bar{X}_2=$	7.5
$s_1=$	0.4	$s_2=$	4.3
$N_1=$	30	$N_2=$	20

14. Again, lonely Tom takes heart in the fact that their GPAs seem higher than the Usaloosa dorm. Are they?

GPAs

Wesaloosa Dorm		Usaloosa Dorm	
$\bar{X}_1=$	3.6	$\bar{X}_2=$	2.9
$s_1=$	0.7	$s_2=$	1.3
$N_1=$	30	$N_2=$	20

15. Finally, Tom realizes that maybe they are spending too much time in the library. Granted, their grades are higher, but they sure are missing out on the dates. The Usaloosa dorm guys are spending only one night per week in the library, compared to the Wesaloosa dorm, which spends an average of 4 nights per week. Is there a significant difference in the time spent in the library between these two populations?

Library Time

Wesaloosa Dorm	Usaloosa Dorm
$P_{s1}=0.57$	$P_{s2}=0.14$
$N_1=$ 340	$N_2=$ 200

DETAILED ANSWERS

1. T(obtained) is 1.12, which does not fall in the critical region, which for the t-distribution is $\pm$ 2.021 (df=N_1+N_2-2, or 57). Therefore, there is no significant difference in income between men and women.

2. This is a one-tailed test in which the research hypothesis would state that Catholics attend more services than Protestants, and would be written as:

$$(H_1 : \mu_1 < \mu_2)$$

Because of the large sample sizes ($N_1 + N_2 > 100$), the Z distribution is used. For a one-tailed test, at the 0.05 level, Z(critical) is -1.65. Z(obtained) is -4.0 and does fall in the critical region. Therefore we reject the null hypothesis of no difference in favor of the research hypothesis. Catholics do attend significantly more religious services than Protestants.

3. Z(obtained) is +.91, which does not fall in the critical region (+1.65). Therefore, we fail to reject the null hypothesis. There is no difference between the GPAs of Sociology majors and that of Psychology majors.

4. Yes, there is a significant difference between the educational achievement of lower income individuals and higher income individuals. At an alpha level of 0.05, t(critical) is: $\pm$2.00 (df=83), whereas t(obtained) is -2.75.

5. Because this is a nominal variable ("Are you afraid of crime?"), you need to use the two-sample proportions equation to solve for Z. In order to do so, you must first solve for P_u, which is 0.70. Next, solve for the standard deviation of the sampling distribution (the denominator of the Z equation), which is 0.05. Finally, solving for Z (obtained), you should get -9.4, which would fall in the critical region of a one tailed test that hypothesize that suburban dwellers are less afraid of crime ($P_{s1} < P_{s2}$).

6. Because the sum of the two samples is less than 100, we use the t-distribution. To find the critical value, find the degrees of freedom, which is $N_1 + N_2 - 2$, or 34. In the t-distribution table, go down the first column to 30 and across to the second to last column (alpha 0.01, two-tailed test). This value (2.750) is the critical value.

7. The degrees of freedom for this problem is 11+12-2, or 21. The critical value for a one-tailed test at the 0.01 level with 21 degrees of freedom is 2.518. Therefore, the test statistic was significant.

8. Z (obtained) is 0.50, which is not significant for a two-tailed, 0.05 test. Therefore, there is no significant difference between Bolivian women and Peruvian women with regards to knowledge of modern contraceptives.

9. Since the combined sample sizes are greater than 100, we would use the Z-distribution. Since the direction of the difference was not hypothesized, we are interested in a two-tailed test at the 0.01 alpha level, which has a critical value of $\pm$2.58.

The Z(obtained) is -1.0, which does not fall in the critical region, therefore, there is no significant difference between these racial groups on the variable Quality of Life.

10. Since both of these samples are large, we use the Z-distribution, with a research hypothesis that Physical Science majors make significantly more money than do the Social Science majors. A one-tailed Z(critical) at the 0.05 level is +1.65. The Z (obtained) is +10.5; therefore, Physical Science majors <u>do</u> make significantly more money after graduation than do Social Science majors.

11. The combined sample size is larger than 100, so we can use the Z-distribution. The Z(obtained) is -1.34, which does not fall within the critical region of the one-tailed test (Z critical=-1.65). Therefore, we fail to reject the null hypothesis. There is no significant difference in the average hours spent studying between those students in the Greek system and those whom are not.

12. Because of the small sample size, we must use the t-distribution. The t(obtained) is 1.52, which suggests that the suburban neighborhoods have more parties than the urban neighborhoods. To determine the significance, though, we must compare this value with the t(critical) value, which is $\pm$2.061. Therefore, there is no significant difference between the number of parties for these two areas.

13. Since the combined sample size is less than 100, we should use the t-distribution. The question asks if there is a difference, not the direction of the difference, thus we should make this a two-tailed test.

With an alpha level of 0.05 and 49 degrees of freedom, the t(critical) is $\pm$2.021. Comparing this to the t(obtained), which is -2.59, we can see that there is a significant difference between the number of dates for these two dorms. Wesaloosa dorm has significantly fewer dates than Usaloosa dorm.

14. Again, because of the small sample size (N_1+N_2 < 100), we must use the t-distribution. The t(obtained) for this problem is +2.41; for a one-tailed test, the t(critical) with 49 degrees of freedom is +1.684. Therefore, there is a significant difference in GPAs between the two dorms: Wesaloosa dorm has a significantly higher GPA than Usaloosa dorm.

15. Wesaloosas spend proportionately more time in the library than the Usaloosas (0.57 compared to 0.14). To determine if it is significantly more so, we need to conduct a two-sample proportions test by finding the Z(obtained) and comparing that to the Z(critical).

The Z(obtained) is +10.75, which falls in the critical region beyond $\pm$1.96. Therefore, the Wesaloosa dorm spends significantly more time in the library than the Usaloosa dorm. Perhaps they should find some dates there.

CHAPTER TEN
HYPOTHESIS TESTING III:
The Analysis of Variance

Analysis of variance, or ANOVA, is very similar in logic to the two-sample t-test (chapter nine). Think back to this test: remember how if we have two samples, let's say males and females, and some interval-ratio variable which we are comparing them on, let's say income, we find the means for each group and compute the t-statistic to find out whether or not these two groups are significantly different from each other. Do men make significantly more income than women, or is there no significant difference?

The t-test is a terrific test **if** you have two samples, but many times you will have more than two samples. If so, then you need to use the ANOVA test. Let's say we are interested in the number of religious services attended per year for different religious groups. If we were to use the t-test, we would have to break everyone into two groups, such as Protestants and non-Protestants, or Catholics and non-Catholics, which is o.k., but there is probably a lot of variety (or **variance**) within the group non-Protestant or non-Catholic. Thus, it would be better to compare multiple groups, such as Protestants, Catholics, Jewish, Other, and None. Since there are now five samples, we can no longer compute the two-sample t-test, therefore we need to use the ANOVA test.

The ANOVA test analyzes the amount of variation (variance) within each group and compares it to the amount of variance between the groups. Two important words to remember are: **within** and **between**; the variation within is compared to the variation between.

Think about the five religious categories above. For a significant difference between these five religious groups to exist, there should be very little variation within each group, but quite a bit of variation between the groups. Generally speaking, then, most Catholics in the sample attend approximately the same number of religious services (small within variation), but attend significantly more or fewer services than at least one of the other groups (large between variation)

For the difference to be significant, then, there should be very little variation within, but quite a bit of variation between groups. If the within variation is small relative to the between variation, then we can reject the null hypothesis, which in this case is that all of the samples will be the same, and is stated as:

$$H_o: \mu_1 = \mu_2 = \mu_3 = \mu_4 = \mu_5$$

The null hypothesis states that all of these populations (Protestants, Catholics, Jewish, Others, and Nones) are all the same on this variable: there is no difference between these groups. The research hypothesis states that at least <u>one</u> population is different and is phrased in sentence structure:

H_1: At lest one of the populations is different.

By stating the research hypothesis this way, we reject the null hypothesis if at least one population is significantly different from all of the others. This means that we are not quite sure which one is different, although looking at the sample means might give us a clue, but that we are fairly certain (95% if the alpha level is 0.05) that <u>at least</u> one group (or population) <u>is</u> different.

The equations for calculating ANOVA look quite daunting, but since the sample sizes are usually fairly small, it is usually just a matter of plug-and-chug.

MULTIPLE-CHOICE QUESTIONS

1. The ANOVA test is often thought of as an extension of the _____________.
   a. Z test
   b. t test
   c. standard deviation
   d. test of one-sample means

2. The ANOVA test is concerned with the difference __________ and __________ groups.
   a. in size, in quantity
   b. in quality, in quantity
   c. between, within
   d. sample means, population means

3. ANOVA is an acronym for:
   a. All Numbers Of Variation Across
   b. Analysis of Numbers OVer Another
   c. ANalysis of Other VAriables
   d. ANalysis Of VAriance

4. SST is an acronym for:
   a. Sum of Squares Total
   b. Summary Statistic Total
   c. Sum of Statistics Together
   d. Squared Summation of T-test

5. SST is equal to _____ + ______.
   a. t(obtained), t(critical)
   b. Z(obtained), Z(critical)
   c. dfw, dfb
   d. SSB, SSW

6. The test statistic computed in ANOVA is the _____________.
a. t-test
b. F ratio
c. A ratio
d. Z test

7. One of the primary differences between the t-test and ANOVA is:
a. ANOVA is used for large samples
b. ANOVA is concerned with only two samples
c. ANOVA can be used for more than two samples
d. there is no difference between the two tests

8. In order to find the critical region, you need ___________ and ___________.
a. degrees of freedom, sample size
b. degrees of freedom, number of categories
c. degrees of freedom within, degrees of freedom between
d. all of the above

9. ANOVA uses the _______ distribution.
a. A
b. F
c. t
d. Z

10. The null hypothesis states:
a. all the sample means are equal
b. the population variances are equal
c. the population means are equal
d. none of the populations are the same

11. The research hypothesis states:
a. all of the sample means are different
b. the population variances are unequal
c. all of the population means is unequal
d. one of the populations means is different

12. The distribution for ANOVA is broken down by ___________ alpha levels.
a. 0.01,0.05, 0.10,0.50,0.95,0.99
b. 0.01, 0.05
c. 0.10, 0.05, 0.01
d. 0.01, 0.95

13. Degrees of freedom within is found with __________ equation.
   a. $N_k - N$
   b. $N_1 + N_2 + N_3 + \ldots N_n - n$
   c. N -1
   d. N -k

14. Degrees of freedom between is found with __________ equation.
   a. k -1
   b. $N_{k1} + N_{k2} + Nk_3 + \ldots N_{kn} - k$
   c. N -1
   d. N -k

15. The test statistic for ANOVA is equal to:
   a. dfb/dfw
   b. dfw/dfb
   c. mean square within/mean square between
   d. mean square between/mean square within

WORK PROBLEMS

1. A researcher collected data on educational attainment in the U.S. Is there a significant difference between the groups?

WHITE	BLACK	HISPANIC
12	12	8
10	12	5
14	11	12
16	14	16
17	16	11
12	12	10
16	13	8
16	10	14
15	16	
18	8	
17		
19		
12		
16		
11		

2. Many counselors suggest taking statistics in either the sophomore or junior year. Below are the average grades for one statistics class. Is there a significant difference between the groups?

FROSH	SOPH	JR.	SR.
79	81	90	87
76	80	93	91
82	87	82	69
91	94	57	82
90	62	62	86
77	90	79	88
78	88	71	71
56	80	74	75
84	76	81	62
87	65	81	54
72	70	85	47
29	69	69	81
	72	97	82
	82	58	77
	81	41	71
	78	91	75
	77	73	76
	79	82	
		84	
		81	

3. At a recent wine festival, three wineries were judged on taste. Below are the scores given to each winery. Is there a significant difference? (Assume the data to be interval-ratio)

WINERY1	WINERY2	WINERY 3
10	9	10
9	9	10
5	10	7
8	8	6
7	7	3
8.5	9	5.5
3	8.5	10
10	6	4.5
9	5	7
5	7.5	8.5
6	7	6
8	9	4
8	8	5
7.5	8	9
6.5	7.5	3

4. GPA of three dorms has been monitored by the administration of a university. Is there cause for worry?

DORM 1	DORM 2	DORM 3
3.5	2.0	2.0
4.0	1.6	4.0
3.0	3.0	3.5
3.0	2.0	3.5
2.7	3.2	3.2
3.1	3.4	3.1
3.2	2.1	3.0
3.3	2.6	1.8
3.5	2.8	2.5
3.2	1.8	2.8
4.0	2.4	3.1
3.8	3.1	2.8
3.6	1.0	2.6
2.8	0.8	2.5

5. Wesaloosa and Usaloosa have been butting heads for quite awhile. Now a third dorm wants to join in. They have collected data on the number of dates per month from a sample of their residents. Is there a significant difference?

WESALOOSA DORM	USALOOSA DORM	ALSALOOSA DORM
4	4	4
1	6	1
0	5	8
5	8	0
2	0	0
1	6	3
3	2	4
4	0	2
1	5	1
2	1	8
0	6	6
4	8	5
3	4	4
8	4	1

6. The number of social events per year for three areas of residence have been collected. Is there a significant difference in the number of social events by area of residence?

RURAL	URBAN	SUBURBAN
0	4	12
3	6	0
1	5	1
0	8	0
1	1	3
5	0	4
12	4	1
	2	6

7. There's been a lot of talk lately about quick weight-loss programs. Is there a significant difference between these groups?

WEIGHT LOSERS	SLIM QUIK	FORGET FAT	NO DISHIN'	POUND NO MORE
5	5	20	2	10
30	10	21	5	14
29	8	16	4	12
26	6	15	7	8
25	2	12	10	4
10	10	8	6	7
8	8	4	7	5
7	7	31	8	22
5	4	15	3	16
22	12	10	2	5

8. Is there a significant difference in the number of calls placed to the police department by area of residence?

INNER CITY	SUB-URBAN	URBAN	RURAL
4	0	7	0
0	7	8	0
8	6	9	5
12	4	11	1
24	12	1	3
52	52	5	2
100	8	0	4
14	18	0	6
87	24	4	0
30	1	6	1
2	6	3	3
1	5	8	4
18	10	1	6

DETAILED ANSWERS

1.

WHITE	BLACK	HISPANIC
12	12	8
10	12	5
14	11	12
16	14	16
17	16	11
12	12	10
16	13	8
16	10	14
15	16	
18	8	
17		
19		
12		
16		
11		

	WHITE	BLACK	HISPANIC
$\sum X_i =$	221	124	84
$\bar{X}_k =$	14.73	12.4	10.5
$\sum X^2 =$	3,361	1,594	970

$\bar{X} = \quad 13.0$

SST $= (3361+1594+970)-(33)(13)^2$
$= 5925-(33)(169)$
$= 5925-5577$
$= 348$

SSB $= 15(14.73-13)^2 + 10(12.4-13)^2 + 8(10.5-13)^2$
$= 15(2.99) + 10(0.36) + 8(6.25)$
$= 44.89 + 3.6 + 50$
$= 98.49$

SSW $= 348 - 98.49$
$= 249.51$

dfw $= N - k = 33 - 3 = 30$
dfb $= k - 1 = 3 - 1 = 2$

MSW $= 249.51/30$
$= 8.32$
MSB $= 98.49/2$
$= 49.25$

F = 49.25/8.32
= 5.92

F(critical) = 3.32
F(obtained) = 5.92

Therefore, we reject the null hypothesis of no difference. At least one of these populations is different than the other on the variable years of education.

2.

FROSH	SOPH	JR.	SR.
79	81	90	87
76	80	93	91
82	87	82	69
91	94	57	82
90	62	62	86
77	90	79	88
78	88	71	71
56	80	74	75
84	76	81	62
87	65	81	54
72	70	85	47
29	69	69	81
	72	97	82
	82	58	77
	81	41	71
	78	91	75
	77	73	76
	79	82	
		84	
		81	
$\sum X_i$=901	1421	1531	1274
$\bar{X}_k$=75.1	83.6	76.6	74.9
$\sum X^2$=70921	113239	120837	97766

$\bar{X}$ = 76.5

SST = $(70921+113239+97766)-(67)(76.5)^2$
= 402763-(67)(5853.25)
= 402763-392100
= 10,663

SSB = $12(75.1-76.5)^2 + 18(83.6-76.5)^2 + 20(76.6-76.5)^2 + 17(74.9-76.5)^2$
= 12(1.96) + 18(50.41) + 20(0.0025) + 17(2.56)

= 23.52 + 907.38 + 0.05 + 111.14
= 1042.36

SSW = 10,663 - 1042.36
= 9,620.64

dfw = N - k = 67 - 4 = 63
dfb = k - 1 = 4 - 1 = 3

MSW = 9620.64/63
= 152.7
MSB = 1042.36/3
= 347.4

F = 347.4/152.7
= 2.27

F(critical) = 2.76
F(obtained) = 2.27

The F(obtained) does not fall in the critical region, therefore we fail to reject the null hypothesis. There is no significant difference between these four groups with regards to grades on the exam.

3.

	WINERY1	WINERY2	WINERY 3
	10	9	10
	9	9	10
	5	10	7
	8	8	6
	7	7	3
	8.5	9	5.5
	3	8.5	10
	10	6	4.5
	9	5	7
	5	7.5	8.5
	6	7	6
	8	9	4
	8	8	5
	7.5	8	9
	6.5	7.5	3
$\sum X_i$=	110.5	118.5	98.5
$\bar{X}_k$=	7.4	7.9	6.6
$\sum X^2$=	832.75	959.75	732.75

$\bar{X} = 7.28$

SST $= (832.75+959.75+732.75)-(45)(7.28)^2$
$= 2525.25-(45)(51.8)$
$= 2525.25-2332.8$
$= 192.45$

SSB $= 15(7.4-7.28)^2 + 15(7.9-7.28)^2 + 15(6.6-7.28)^2$
$= 15(0.0144) + 15(0.3844) + 15(0.4624)$
$= 0.216 + 5.766 + 6.936$
$= 12.92$

SSW $= 192.45 - 12.92$
$= 179.53$

dfw $= N - k = 45 - 3 = 42$
dfb $= k - 1 = 3 - 1 = 2$

MSW $= 179.53/42$
$= 4.27$
MSB $= 12.92/2$
$= 6.46$

F $= 6.46/4.27$
$= 1.51$

F(critical) = 3.23
F(obtained) = 1.51

Therefore, we fail to reject the null hypothesis. There is no significant difference between the ratings of these three wineries.

4.

DORM 1	DORM 2	DORM 3
3.5	2.0	2.0
4.0	1.6	4.0
3.0	3.0	3.5
3.0	2.0	3.5
2.7	3.2	3.2
3.1	3.4	3.1
3.2	2.1	3.0
3.3	2.6	1.8
3.5	2.8	2.5
3.2	1.8	2.8
4.0	2.4	3.1
3.8	3.1	2.8

	3.6	1.0	2.6
	2.8	0.8	2.5
$\sum X_i =$	46.7	31.8	40.4
$\bar{X}_k =$	3.34	2.27	2.86
$\sum X^2 =$	158.01	76.66	121.14

$\bar{X} =$ 2.83

SST = $(158.01+79.66+121.14)-(42)(2.83)^2$
= $358.81-(42)(8.01)$
= $358.81-336.42$
= 22.39

SSB = $14(3.34-2.83)^2 + 14(2.27-2.83)^2 + 14(2.86-2.83)^2$
= $14(0.26) + 14(0.31) + 14(0.0009)$
= $3.64 + 4.39 + 0.0126$
= 7.86

SSW = $22.39 - 7.86$
= 14.53

dfw = $N - k = 42 - 3 = 39$
dfb = $k - 1 = 3 - 1 = 2$

MSW = $14.53/39$
= 0.37
MSB = $7.86/2$
= 3.93

F = $3.93/0.37$
= 10.62

F(critical) = 3.32
F(obtained) = 10.62

The obtained value falls in the critical region, therefore we reject the null hypothesis of no difference. At least one of these dorms is significantly different than the other dorms.

5.

	WESALOOSA DORM	USALOOSA DORM	ALSALOOSA DORM
	4	4	4
	1	6	1
	0	5	8
	5	8	0
	2	0	0
	1	6	3
	3	2	4
	4	0	2
	1	5	1
	2	1	8
	0	6	6
	4	8	5
	3	4	4
	8	4	1
$\sum X_i=$	38	69	57
$\bar{X}_k=$	2.71	4.93	4.07
$\sum X^2=$	166	439	369

$\bar{X} =$ 3.90

SST = $(166+439+369)-(42)(3.90)^2$
= $974-(42)(15.21)$
= $974-638.82$
= 335.18

SSB = $14(2.71-3.90)^2 + 14(4.93-3.90)^2 + 14(4.07-3.90)^2$
= $14(1.42) + 14(1.06) + 14(0.03)$
= $19.83 + 14.85 + 0.40$
= 35.08

SSW = $335.18 - 35.08$
= 300.10

dfw = $N - k = 42 - 3 = 39$
dfb = $k - 1 = 3 - 1 = 2$

MSW = $300.10/39$
= 7.69
MSB = $35.08/2$
= 17.54
F = $17.54/7.69$
= 2.28

F(critical) = 3.32
F(obtained) = 2.28

We fail to reject the null hypothesis. There is no significant difference in the number of dates between the dorms. Therefore, although Wesaloosa dorm has fewer dates than either Usaloosa or Alsaloosa, the difference is not significant.

6.

	RURAL	URBAN	SUBURBAN
	0	4	12
	3	6	0
	1	5	1
	0	8	0
	1	1	3
	5	0	4
	12	4	1
		2	6
$\sum X_i =$	22	34	35
$\bar{X}_k =$	3.14	3.4	3.5
$\sum X^2 =$	180	172	257

$\bar{X} =$ 3.37

SST $= (180+172+257)-(27)(3.37)^2$
$= 609-(27)(11.36)$
$= 609-306.72$
$= 302.28$

SSB $= 7(3.14-3.37)^2 + 10(3.4-3.37)^2 + 10(3.5-3.37)^2$
$= 7(0.05) + 10(0.0009) + 10(0.0169)$
$= 0.37 + 0.009 + 0.169$
$= 0.548$

SSW $= 302.28 - 0.548$
$= 301.812$

dfw $= N - k = 27 - 3 = 24$
dfb $= k - 1 = 3 - 1 = 2$

MSW $= 302.28/24$
$= 12.6$
MSB $= 0.548/2$
$= 0.274$

F = 0.274/12.6
= 0.022

F(critical) = 3.35
F(obtained) = 0.022

Therefore, we fail to reject the null hypothesis. There is no significant difference between these three areas of residence with regards to the number of social events.

7.

	WEIGHT LOSERS	SLIM QUIK	FORGET FAT	NO DISHIN'	POUND NO MORE
	5	5	20	2	10
	30	10	21	5	14
	29	8	16	4	12
	26	6	15	7	8
	25	2	12	10	4
	10	10	8	6	7
	8	8	4	7	5
	7	7	31	8	22
	5	4	15	3	16
	22	12	10	2	5
$\sum X_i =$	162	72	152	54	103
$\bar{X}_k =$	16.7	7.2	15.2	5.4	10.3
$\sum X^2 =$	3789	609	2832	346 1359	

$\bar{X} =$ 10.96

SST = $(3789+602+2832+346+1359)-(50)(10.96)^2$
= 8928-(50)(120.12)
= 8928-6006.1
= 2921.9

SSB = $10(16.7-10.96)^2 + 10(7.2-10.96)^2 + 10(15.2-10.96)^2 + 10(10.3-10.96)^2$
= 10(32.95) + 10(14.14) + 10(17.98) + 10(30.91) + 10(0.436)
= 329.5 + 141.4 + 179.8 + 309.1 + 4.36
= 964.16

SSW = 2,921.9 - 964.16
= 1,957.74

dfw = N - k = 50 - 5 = 45
dfb = k - 1 = 5 - 1 = 4

MSW = 1957.74/45
= 43.51
MSB = 964.12/4
= 241.03

F = 241.03/43.51
= 5.54
F(critical) = 2.61
F(obtained) = 5.54

Therefore, we reject the null hypothesis. There is a significant difference in the average weight lost by weight-loss program.

8.

	INNER CITY	URBAN	SUB-URBAN	RURAL
	4	0	7	0
	0	7	8	0
	8	6	9	5
	12	4	11	1
	24	12	1	3
	52	52	5	2
	100	8	0	4
	14	18	0	6
	87	24	4	0
	30	1	6	1
	2	6	3	3
	1	5	8	4
	18	10	1	6
$\sum X_i=$	352	153	63	39
$\bar{X}_k=$	27.1	11.77	4.85	3.0
$\sum X^2=$	22498	4075	467	153

$\bar{X} =$ 11.67

SST $= (22498+4075+467+153)-(52)(11.67)^2$
$= 27193-(52)(136.19)$
$= 27193-7081.9$
$= 20111.1$

SSB $= 13(27.1-11.67)^2 + 13(11.77-11.67)^2 + 13(4.85-11.67)^2 + 13(3.0-11.67)^2$
$= 13(238.1) + 13(0.01) + 13(46.51) + 13(75.17)$
$= 3095.3 + 0.13 + 604.63 + 977.21$
$= 4677.27$

SSW = 20111.1 - 4677.27
= 15433.83

dfw = N - k = 52 - 4 = 48
dfb = k - 1 = 4 - 1 = 3

MSW = 15433.83/48
= 321.54

MSB = 4677.27/3
= 1559.09

F = 1559.09/321.54
= 4.85

F(critical) = 2.84
F(obtained) = 4.85

Therefore we reject the null hypothesis of no difference. There is a significant difference in the number of calls placed to police between these places of residence.

CHAPTER ELEVEN
HYPOTHESIS TESTING IV:
Nonparametric Tests for Variables Measured at the Ordinal Level

Non-parametric refers to tests in which the shape of the distribution is not assumed to be normal. If you think back to the last three chapters, hypothesis testing for one- and two-samples, and ANOVA, we made the assumption that the shape of the population distribution was normal, thus allowing us to make inferences about the population from our sample. In non-parametric tests, though, this assumption is not made. In fact, non-parametric tests are used whenever the assumption of normality can not be met.

Two non-parametric tests for ordinal data are presented in this chapter: (1) the Mann-Whitney U and (2) the runs test. In both tests, you have two independent groups, such as males and females, in which you are determining whether or not these groups are significantly different from one another on some specific **ordinal** variable.

If you think back to ordinal-level variables, you should remember that these variables can be rank-ordered, but the distance between two scores really doesn't mean anything. Instead, all we can tell is that one score is higher than another, not the distance between the two.

In these two tests of significance, each group is ordered, from highest to lowest, then a **rank** is assigned to each score. Let's say Tom, Robert, and Jim have the following prestige scores:

	Score
Robert	86
Tom	89
Jim	85

If we were to rank these three scores, Tom would receive the rank of one, Robert would receive two, and Jim would receive the rank of three. In this same manner, these two tests rank the groups.

In the Mann-Whitney U, the two groups are ranked together. Thus, if we were interested in prestige between males and females, the male or female with the highest score would receive the rank of one, and so on. Remember: both groups are ranked together, so that in the end, you should have the same number of rankings as the **combined** sample size.

The runs test uses the same logic, but instead of ranking each case, we find any sequence of scores (or "runs") and count the number of total runs. For example, let's say we have the following data on job satisfaction for both managers and non-managers:

Job Satisfaction

Managers Case	Score	Non-Managers Case	Score
Bob	20	Tom	21
John	19	Joe	14
Sue	15	Donna	12
Ruth	7	Sheryl	9

The first run would start with the highest score, Tom (21). Since Joe does not have the next highest score (Bob does), the run ends there. Runs must be within each category--not across categories! The second run would begin with Bob (20), and include John and Sue. The third run would be Donna (12) and Sheryl (9), and the fourth run would be Ruth (7). Therefore, there are four runs in this data. If the null hypothesis is true, that there is no difference between these two groups, then there should be many runs. If the null hypothesis is false, then there should be very few runs.

Remember that both of these tests are for ordinal data when you cannot make an assumption about the shape of the distribution. If you have interval-ratio data and/or can assume a normal distribution, it is much better to use one of the two-sample means or proportions tests that we learned about in Chapter nine.

MULTIPLE-CHOICE QUESTIONS

1. The term "non-parametric" means:
   a. no parameters are set for the population value
   b. the sampling distribution is set to be equal to zero
   c. the means are used to estimate the parameter sets
   d. no assumptions are made about the shape of the distribution

2. Mann-Whitney U is a(n) ______________ test.
   a. interval-ratio level
   b. normal distribution
   c. ordinal level
   d. two-sample means

3. The Mann-Whitney U test uses ________ to test for ___________.
   a. interval-ratio data, normality
   b. normally distributed data, significance
   c. sample means, significance
   d. ordinal data, significance

4. What two assumptions are made when using non-parametric tests?
   a. sampling distribution is normal and population variances are equal
   b. independent random samples and level of measurement is ordinal
   c. sample means are equal to the population means and large sample size
   d. all of the above

5. When finding U, the _______ value is always used.
   a. lower
   b. higher
   c. middle
   d. upper

6. If you have many ties across the two samples it is best to use ____________ test.
   a. two-sample means
   b. Mann-Whitney U
   c. runs test
   d. do not use any test

7. The null hypothesis in the runs test will always be:
   a. no significant difference between the populations
   b. no significant difference between the population means
   c. no significant difference between the variances of the two populations
   d. there is no null hypothesis

8. A **runs** is defined as:
   a. the speediest calculation
   b. any sequence of scores
   c. any score not in the order expected
   d. any one score

9. If the null hypothesis is true, there should be ________ runs.
   a. very few
   b. some
   c. none
   d. many

10. If the null hypothesis is false, there should be ________ runs.
    a. very few
    b. some
    c. none
    d. many

WORK PROBLEMS

1. A scale measuring quality of life was administered to two groups, whites and blacks, in which high scores indicated a high quality of life, low scores indicated a low quality of life. The following data were collected. Is there a significant difference between these two groups with regards to their quality of life?

Quality of Life

White		Black	
Case	Score	Case	Score
1	95	1	99
2	94	2	94
3	91	3	93
4	90	4	90
5	87	5	90
6	84	6	88
7	82	7	84
8	78	8	80
9	78	9	72
10	72	10	70
11	61	11	70
12	60	12	68
13	58	13	67
14	53	14	65
15	53	15	58
16	47	16	55
17	43	17	48
18	40	18	42
19	37	19	40
20	35	20	39

2. Occupational prestige was measured for both males and females. If high scores indicate higher prestige, do males have significantly higher prestige than females?

Prestige

Male Case	Male Score	Female Case	Female Score
1	89	1	83
2	84	2	80
3	78	3	78
4	70	4	70
5	64	5	70
6	60	6	65
7	58	7	52
8	56	8	43
9	42	9	41
10	40	10	35
11	35	11	22

3. A researcher hypothesizes that college graduates will have fathers with higher prestige scores than the fathers of people with high school diplomas. She collected the following data. Are college graduates' fathers more likely to have higher prestige than high school graduates?

Fathers Prestige

College Grads Case	College Grads Score	High School Grads Case	High School Grads Score
1	89	1	78
2	88	2	67
3	82	3	66
4	79	4	65
5	77	5	64
6	70	6	58
7	69	7	57
8	63	8	56
9	62	9	53
10	61	10	51
11	60	11	49
12	59	12	47
13	55	13	44
14	54	14	43
15	46	15	38
16	45	16	36
17	42	17	35
18	41	18	32
19	40	19	30
20	39	20	29

4. An urban sociologist hypothesized that people living in the city would be happier than people living in the suburbs. He collected data for both groups using a life satisfaction scale (high scores indicate higher life satisfaction). Is there a significant difference?

Life Satisfaction

Urban Residents		Suburban Residents	
Case	Score	Case	Score
1	65	1	60
2	65	2	56
3	63	3	50
4	58	4	48
5	52	5	40
6	47	6	40
7	45	7	38
8	45	8	35
9	40	9	29
10	38	10	26
11	37	11	20
12	30	12	15

DETAILED ANSWERS

1. Because of the number of ties, you probably want to use the Mann-Whitney U. Begin by ranking each case across both categories:

Quality of Life

White			Black		
Case	Score	Rank	Case	Score	Rank
1	95	2	1	99	1
2	94	3.5	2	94	3.5
3	91	6	3	93	5
4	90	8	4	90	8
5	87	11	5	90	8
6	84	12.5	6	88	10
7	82	14	7	84	12.5
8	78	16.5	8	80	15
9	78	16.5	9	72	18.5
10	72	18.5	10	70	20.5
11	61	25	11	70	20.5
12	60	26	12	68	22
13	58	27.5	13	67	23
14	53	30.5	14	65	24
15	53	30.5	15	58	27.5
16	47	33	16	55	29

17	43	34	17	48	32
18	40	36.5	18	42	35
19	37	39	19	40	36.5
20	35	40	20	39	48
		$\sum R_1 = 430.5$			$\sum R_2 = 389.5$

$$U = N_1 N_2 + \frac{N_1(N_1 - 1)}{2} - \Sigma R_1$$

$$U = 20(20) + \frac{20(20-1)}{2} - 389.7 = 200.3$$

$$\mu_u = \frac{N_1 N_2}{2} = \frac{(20)(20)}{2} = 200$$

$$\sigma_u = \sqrt{\frac{N_1 N_2 (N_1 + N_2 + 1)}{12}}$$

$$\sigma_u = \sqrt{\frac{(20)(20)(20+20+1)}{12}} = 36.97$$

$$Z = \frac{U - \mu_u}{\sigma_u} = \frac{200.3 - 200}{36.97} = 0.008$$

Therefore, we found the Z(obtained) to be 0.008, which does not fall in the critical region (Z critical = $\pm$1.96). Therefore we fail to reject the null hypothesis of no difference. There is no significant difference in quality of life between blacks and whites.

2.

Prestige

Male Case	Score	Rank	Female Case	Score	Rank
1	89	1	1	83	3
2	84	2	2	80	4
3	78	5.5	3	78	5.5
4	70	8	4	70	8
5	64	11	5	70	8
6	60	12	6	65	10
7	58	13	7	52	15
8	56	14	8	43	16
9	42	17	9	41	18
10	40	19	10	35	20.5
11	35	20.5	11	22	22
		$\sum R_1=107$			$\sum R_2=130$

$$U=N_1N_2+\frac{N_1(N_1-1)}{2}-\sum R_1$$

$$U=11(11)+\frac{11(11-1)}{2}-107=69$$

$$\mu_u=\frac{N_1N_2}{2}=\frac{(11)(11)}{2}=60.5$$

$$\sigma_u=\sqrt{\frac{N_1N_2(N_1+N_2+1)}{12}}$$

$$\sigma_u=\sqrt{\frac{(11)(11)(11+11+1)}{12}}=15.23$$

$$Z = \frac{U - \mu_u}{\sigma_u} = \frac{69-60.5}{15.23} = 0.56$$

The critical value for a one-tailed test is +1.65. The Z(obtained) of 0.56 does not fall in this region, therefore we fail to reject the null hypothesis.

3. Because there are no ties, we can compute a runs test. To do so, start with the highest number (case 1 of the college grads) and find all the continuous cases in that category: this will be run one. Do it for all groups of cases.

Fathers Prestige

College Grads Case	Score	Run	High School Grads Case	Score	Run
1	89 ┐		1	78]	2
2	88	1	2	67 ┐	
3	82		3	66	4
4	79 ┘		4	65	
5	77 ┐		5	64 ┘	
6	70	3	6	58 ┐	
7	69 ┘		7	57	6
8	63 ┐		8	56 ┘	
9	62		9	53 ┐	
10	61	5	10	51	8
11	60		11	49	
12	59 ┘		12	47 ┘	
13	55 ┐	7	13	44 ┐	
14	54 ┘		14	43 ┘	10
15	46 ┐		15	38 ┐	
16	45 ┘	9	16	36	
17	42 ┐		17	35	
18	41	11	18	32	
19	40		19	30	12
20	39 ┘		20	29 ┘	

$$\mu_R = \frac{2N_1 N_2}{N_1 + N_2} + 1$$

$$\mu_R = \frac{2(20)(20)}{20+20} + 1 = 21$$

$$\sigma_R = \sqrt{\frac{2N_1N_2(2N_1N_2 - N_1 - N_2)}{12}}$$

$$\sigma_R = \sqrt{\frac{2[(20)(20)][(2(20)(20) - (20) - (20)]}{12}} = 3.04$$

$$Z(obt) = \frac{R - \mu_R}{\sigma_R}$$

$$Z(obt) = \frac{12 - 21}{3.04} = -2.96$$

The Z(obtained) of -2.96 does fall in the critical region, therefore we reject the null hypothesis. College graduates have fathers with significantly higher prestige than high school graduates.

4.

Life Satisfaction

Urban Residents			Suburban Residents		
Case	Score	Rank	Case	Score	Rank
1	65	1.5	1	60	4
2	65	1.5	2	56	6
3	63	3	3	50	8
4	58	5	4	48	9
5	52	7	5	40	14
6	47	10	6	40	14
7	45	11.5	7	38	16.5
8	45	11.5	8	35	19
9	40	14	9	29	21
10	38	16.5	10	26	22
11	37	18	11	20	23
12	30	20	12	15	24
		$\sum R_1 = 119.5$			$\sum R_2 = 180.5$

$$U=N_1N_2+\frac{N_1(N_1-1)}{2}-\Sigma R_1$$

$$U=12(12)+\frac{12(12-1)}{2}-119.5=102.5$$

$$\mu_u=\frac{N_1N_2}{2}=\frac{(12)(12)}{2}=72$$

$$\sigma_u=\sqrt{\frac{N_1N_2(N_1+N_2+1)}{12}}$$

$$\sigma_u=\sqrt{\frac{(12)(12)(12+12+1)}{12}}=12.49$$

$$Z=\frac{U-\mu_u}{\sigma_u}=\frac{102.5-72}{12.49}=+2.44$$

The Z(obtained) of +2.44 falls outside the critical value (Z critical) of +1.96. Therefore, we reject the null hypothesis. Urban residents are significantly more satisfied with life than suburban residents.

CHAPTER TWELVE
HYPOTHESIS TESTING V:
Chi Square

One of the most commonly used statistics in the social sciences is the chi (pronounced "kai") square test of independence. Like the measures discussed in chapter ten, chi square is also a **non-parametric** statistic, which means that no assumptions are made about the shape of the distribution. This makes this little statistic a very popular one indeed. Another reason for its popularity is its ability to be used with nominal data.

Chi square is especially easy to use when you have small sample sizes and data in bivariate tables (data arranged in a table format, so that the groups, such as males and females, are along the top and the categories for the variable are along the side). In each cell, then, are the frequencies for that category on that value of the variable.

In the chi square test, the null hypothesis is that the two variables are **independent** of one another, meaning that being placed in one of categories of the first variable is independent of being in either category of the second variable. Think of it like this: if we have two groups, males and females, and the variable attitude toward abortion rights (pro-choice or pro-life), we can say that the two variables (gender and attitude toward abortion rights) are independent if the chance of being in either category of either pro-choice or pro-life is not affected by gender. Said another way: being male or female would not affect the chance of being either pro-choice or pro-life. If the two variables were **dependent** then being either male or female would affect being either pro-choice or pro-life.

Therefore, the null hypothesis is always that the two variables (gender and attitude toward abortion rights) are independent of one another. The research hypothesis, then, is that they are dependent on one another.

The chi square statistic is affected by sample size; in fact, the chi square statistic increases at the same rate as the sample size. Thus, if you have a large sample very small differences will be exaggerated in this statistic. Use caution, also, when you have any cells with frequencies less than five. The chi square statistic cannot be adequately computed with such small cell sizes, although you can correct for this problem on 2x2 tables with Yate's correction for continuity (see section 11.7 in your textbook).

The chi square test uses the row and column marginals to create **expected frequencies**, or the cell frequencies we would expect to get <u>if</u> the two variables were independent. These expected frequencies are then compared to the **observed frequencies**, or the cell frequencies that are actually observed with our data. If there is a large difference between the expected frequencies and the observed frequencies, then we conclude that the two variables are dependent on one another and we reject the null hypothesis.

MULTIPLE-CHOICE QUESTIONS

1. In the null hypothesis statement for the chi square test, the two variables are hypothesized to be:
   a. dependent
   b. independent
   c. associated
   d. equal variances

2. Row marginals are:
   a. the total sample size
   b. the vertical subtotals
   c. the horizontal subtotals
   d. the largest cell frequency

3. Observed frequencies are:
   a. cell frequencies expected if the two variables are independent
   b. total frequencies observed
   c. total frequencies expected
   d. cell frequencies observed from the raw data

4. Expected frequencies are:
   a. cell frequencies expected if the two variables are independent
   b. total frequencies observed
   c. total frequencies expected
   d. cell frequencies observed from the raw data

5. ________ frequencies are the raw data, whereas _________ frequencies are found with row and column marginals.
   a. Expected, observed
   b. Total, expected
   c. Observed, total
   d. Observed, expected

6. In order to find the observed frequencies:
   a. multiply the row and column marginals and divide by the sample size
   b. add up all of the frequencies for each cell from the raw data
   c. divide the cell frequencies by the total sample size
   d. multiply the sample size by each cell frequency

7. In order to find the expected frequencies:
   a. multiply the row and column marginals and divide by the sample size
   b. add up all of the frequencies for each cell from the raw data
   c. divide the cell frequencies by the total sample size
   d. multiply the sample size by each cell frequency

8. In order to test for independence, _______________ is/are compared to _______________.
   a. chi square obtained, chi square critical
   b. row marginals, column marginals
   c. degrees of freedom, cell frequencies
   d. marginal totals, sample size

9. Degrees of freedom are found with which of the following formulas?
   a. (number of rows) (sample size)
   b. (number of rows)(number of columns)
   c. (number of rows-1)(number of columns-1)
   d. (sample size, N) (number of cells)

10. In order to find the critical value in the chi square test, you need __________ and __________.
    a. row marginals, sample size
    b. degrees of freedom, critical value
    c. sample size, column marginals
    d. degrees of freedom, alpha level

11. The chi square test is affected by __________.
    a. row marginals
    b. sample size
    c. cell frequencies
    d. all of the above

12. The chi square critical value for a five by four matrix, and an alpha level of 0.01 is:
    a. 3.571
    b. 12
    c. 21.026
    d. 26.217

13. If a chi square test statistic was calculated to be 4.72, with two degrees of freedom, it would be rejected at ________ alpha level.
    a. 0.001
    b. 0.01
    c. 0.05
    d. 0.10

14. In order to reject a null hypothesis, the obtained value must fall _________ the critical value.
    a. within
    b. to the left
    c. beyond
    d. nowhere near

15. The sampling distribution of the chi square is ___________ .
a. positively skewed
b. negatively skewed
c. normal in shape
d. determined by the means of the population

WORK PROBLEMS

1. The following data show the distribution of job satisfaction and life satisfaction. Using only the one expected frequency, can you find the other three with subtraction?

F_o

Life Satisfaction	Job Satisfaction: High	Job Satisfaction: Low	
High	54	29	83
Low	31	65	96
	85	94	179

F_e

Life Satisfaction	Job Satisfaction: High	Job Satisfaction: Low	
High	39.4		83
Low	_	_	96
	85	94	179

2. From the data above, what can we say about the two variables, job satisfaction and life satisfaction? Are the two variables independent of one another?

3. For the data below, find the expected frequencies using the subtraction method and only those cell frequencies given. Note: since this is a two by three matrix, you need only two pieces of information to complete the matrix, therefore the degrees of freedom is two.

F_e

	Dorm 1	Dorm Dorm 2	Dorm 3	
Favor	17.9			51
Oppose	__	14.7	__	49
	35	30	35	100

4. The following data were collected to determine if gender and marital happiness were related. Are the two variables independent of one another?

F_o

Marital Satisfaction	Gender Male	Female	
Very Happy	30	45	75
Moderately Happy	50	45	95
Unhappy	20	10	30
	100	100	200

5. A researcher suspected that attitudes toward gun control were dependent on gender. Does the data below support his hypothesis?

F_o

AttitudeToward Gun Control	Gender Male	Female	
Favor	45	42	87
Oppose	39	49	88
	84	91	175

6. Knowledge of birth control between men and women were measured. Are the two variables independent of one another? What problems do we encounter in this test?

F_o

Knowledge of Birth Control	Gender Male	Female	
High	175	430	605
Moderate	512	220	732
Low	215	160	375
	902	810	1710

7. With more women in the work force these days, a sociologist wondered whether men were taking over more household responsibilities. She collected the following data. Are gender and housework independent?

F_o

Amount of Housework	Gender Male	Female	
All	15	46	61
Most	20	53	73
Some	69	27	96
None	11	9	20
	115	135	250

8. Crime is perceived to be a greater threat in the city than in the suburbs. A criminologist collected data on the victimization of crimes for both areas. Do the data below indicate that the two variables are dependent?

F_o

Victim of Crime	Area of Residence Urban	Suburban	
Yes	40	25	65
No	25	35	60
	65	60	125

9. What is the critical value for a chi square test with five degrees of freedom at the 0.01 level?

10. If a researcher found a chi square obtained of 6.75, in a test with three degrees of freedom, would she be able to reject the null hypothesis at the 0.01 level? If not, at what level could she reject the null hypothesis?

11. In a poll for the approval rating of a new Democratic president, a newspaper showed the following data. Is the approval rating dependent on political party preference?

F_o Approval of President	Political Party Preference: Republican	Democrat	
Yes	14	29	43
No	24	11	35
	38	40	78

12. In a similar poll, respondents were asked to rate their politics as either liberal or conservative. Are these labels independent of political party?

F_o Political Party Affiliation	Political Ideology: Liberal	Conservative	
Republican	3	25	28
Democrat	29	2	31
Independent	10	8	18
	42	35	77

13. A newly elected congressman is considering introducing a new bill to legalize marijuana, but first wants to know if gender plays a role in the attitude toward the issue. Are the two variables independent?

F_o Legalization of Marijuana	Gender: Male	Female	
Approve	10	6	16
Disapprove	5	9	14
No Opinion	9	11	20
	24	26	50

14. Bob Jherque, the new mayor, has proposed the construction of a new water plant. Is the approval rating dependent on whether you live in town or on a farm?

F_o

Approval of Water Plant	Place of Residence: Town	Farm	
Yes	33	46	79
No	37	30	67
	70	76	146

DETAILED ANSWERS

1.

F_e

Life Satisfaction	Job Satisfaction: High	Low	
High	39.4	43.6	83
Low	45.6	50.4	96
	85	94	179

2.

$$\chi^2 = \frac{(54-39.4)^2}{39.4} + \frac{(29-43.6)^2}{43.5} + \frac{(31-45.6)^2}{45.6} + \frac{(65-50.4)^2}{50.4}$$

$$\chi^2 = 5.4 + 4.9 + 4.7 + 4.2 = 19.2$$

χ^2 critical with one degree of freedom is 3.841. The χ^2 obtained is 19.2. Therefore we reject the null hypothesis of independence--job satisfaction and life satisfaction are dependent on one another.

3.
F_e

	Dorm 1	Dorm 2	Dorm 3	
Favor	17.9	15.3	17.8	51
Oppose	17.1	14.7	17.2	49
	35	30	35	100

4.
F_e

Marital Satisfaction	Gender: Male	Female	
Very Happy	37.5	37.5	75
Moderately Happy	47.5	47.5	95
Unhappy	15	15	30
	100	100	200

$$\frac{0-37.5)^2}{37.5}+\frac{(45-37.5)^2}{37.5}+\frac{(50-47.5)^2}{47.5}+\frac{(45-47.5)^2}{47.5}+\frac{(20-15)^2}{15}+\frac{(}{}$$

$$\chi^2=1.5+1.5+0.13+0.13+1.7+1.7=6.66$$

χ^2 critical with two degrees of freedom is 5.991. The χ^2 obtained is 6.66. Therefore we reject the null hypothesis of independence--marital satisfaction and gender are dependent on one another.

5.
F_e

Attitude Toward Gun Control	Gender: Male	Female	
Favor	41.8	45.2	87
Oppose	42.2	45.8	88
	84	91	175

$$\chi^2=\frac{(45-41.8)^2}{41.8}+\frac{(42-45.2)^2}{45.2}+\frac{(39-42.2)^2}{42.2}+\frac{(49-45.8)^2}{45.8}$$

$$\chi^2=0.25+0.23+0.25+0.23=0.96$$

χ^2 critical with one degree of freedom is 3.841. The χ^2 obtained is 0.96. Therefore we fail to reject the null hypothesis of independence--attitude toward gun control and gender are independent of one another.

6.

F_e

Knowledge of Birth Control	Gender: Male	Female	
High	318.8	286.2	605
Moderate	285.7	346.3	732
Low	197.6	177.4	375

$$\chi^2=\frac{(175-318.8)^2}{318.8}+\frac{(512-385.7)^2}{385.7}+\frac{(215-197.6)^2}{197.6}$$

$$+\frac{(430-286.2)^2}{286.2}+\frac{(220-346.3)^2}{346.3}+\frac{(160-177.4)^2}{177.4}$$

$$\chi^2=64.8+41.4+1.5+72.2+46.1+1.7=227.7$$

With two degrees of freedom, at a 0.05 alpha level, the critical value is 5.991. The obtained value is 227.7, which definitely falls in the critical region. One of the problems, though, is the large sample sizes. Because the sample sizes are 902 and 810, even the smallest differences will appear significant. Therefore, although we reject the null hypothesis of independence, we do so with caution.

7.

F_e

Amount of Housework	Gender: Male	Female	
All	28.1	32.9	61
Most	33.6	39.4	73
Some	44.2	51.8	96
None	9.2	10.8	20
	115	135	250

$$\chi^2 = \frac{(15-28.1)^2}{28.1} + \frac{(20-33.6)^2}{33.6} + \frac{(69-44.2)^2}{44.2} + \frac{(11-9.2)^2}{9.2}$$

$$+ \frac{(46-32.9)^2}{32.9} + \frac{(53-39.4)^2}{39.4} + \frac{(27-51.8)^2}{51.8} + \frac{(9-10.8)^2}{10.8}$$

$$\chi^2 = 6.1 + 5.5 + 13.9 + 0.4 + 5.2 + 4.7 + 11.8 + 0.3 = 47.9$$

With three degrees of freedom, the critical value is 7.815. The obtained value (47.8) falls in this region, therefore we can conclude that gender and amount of housework are dependent on one another.

8.

F_e

Victim of Crime	Area of Residence: Urban	Suburban	
Yes	33.8	31.2	65
No	31.2	28.8	60
	65	60	125

$$\chi^2 = \frac{(40-33.8)^2}{33.8} + \frac{(25-31.2)^2}{31.2} + \frac{(25-31.2)^2}{31.2} + \frac{(25-28.8)^2}{28.8}$$

The chi square obtained falls in the critical region, therefore we reject the null hypothesis of independence. The variables area of residence and victim of crime are dependent on each other.

$$\chi^2 = 1.14 + 1.23 + 1.23 + 1.33 = 4.93$$

9. The critical value for a chi square test with five degrees of freedom at the 0.01 level is 15.086.

10. The critical value for a chi square test at the 0.01 level, with three degrees of freedom is 11.341, so she could not reject the null hypothesis at that level, but could reject at the 0.10 level, which would indicate the probability of error is 10% rather than 1%.

11.

F_e

Approval of President	Political Party Preference: Republican	Democrat	
Yes	20.9	22.1	43
No	17.1	17.9	35
	38	40	78

$$\chi^2 = \frac{(14-20.9)^2}{20.9} + \frac{(29-22.1)^2}{22.1} + \frac{(24-17.1)^2}{17.1} + \frac{(11-17.9)^2}{17.9}$$

$$\chi^2 = 2.28 + 2.15 + 2.78 + 2.66 = 9.87$$

With one degree of freedom, the χ^2(critical) is 3.841 at the 0.05 level, therefore we reject the null hypothesis of independence. Approval of the President is dependent on political party preference.

12.

F_o

Political Party Affiliation	Political Ideology: Liberal	Conservative	
Republican	15.3	12.7	28

Democrat	16.9	14.1	31
Independent	9.8	8.2	18
	42	35	77

$$\chi^2 = \frac{(3-15.3)^2}{15.3} + \frac{(25-12.7)^2}{12.7} + \frac{(29-16.9)^2}{16.9} + \frac{(2-14.1)^2}{14.1} +$$

$$+ \frac{(10-9.8)^2}{9.8} + \frac{(8-8.2)^2}{8.2} = 40.9$$

The obtained value falls in the critical region. Therefore, we reject the null hypothesis of independence in favor of the research hypothesis. Political ideology is dependent on political party.

13.

F_e

Legalization of Marijuana	Gender		
	Male	Female	
Approve	7.7	8.3	16
Disapprove	6.7	7.3	14
No Opinion	9.6	10.4	20
	24	26	50

$$\frac{(10-7.7)^2}{7.7} + \frac{(6-8.3)^2}{8.3} + \frac{(5-6.7)^2}{6.7} + \frac{(9-7.3)^2}{7.3} + \frac{(9-9.6)^2}{9.6} + \frac{(11-10}{10.}$$

$$\chi^2 = 0.69 + 0.64 + 0.43 + 0.396 + 0.038 + 0.035 = 2.229$$

With two degrees of freedom, we fail to reject the null hypothesis at the 0.05 level. Gender and attitude toward legalizing marijuana are not dependent on each other.

14.

F_e

Approval of Water Plant	Place of Residence Town	Farm	
Yes	37.9	41.1	79
No	32.1	34.9	67
	70	76	146

$$X^2 = \frac{(33-37.9)^2}{37.9} + \frac{(46-41.1)^2}{41.1} + \frac{(37-32.1)^2}{32.1} + \frac{(30-34.9)^2}{34.9}$$

$$X^2 = 0.63 + 0.58 + 0.75 + 0.69 = 2.65$$

We fail to reject the null hypothesis of independence. Place of resident is independent of approval of the water plant.

CHAPTER THIRTEEN
BIVARIATE ASSOCIATION:
Introduction & Basic Concepts

Bivariate means simply that: bi, or **two**, variables. When we are interested in the relationship between two variables, especially when we are concerned with the association, then we are talking about a bivariate relationship.

When we look at the relationship between two variables, we need to know which variable is the cause and which is effect. These are also known as the **independent** and **dependent** variables. The independent variable is the one hypothesized to cause a reaction or an effect in the dependent variable. Think about the following example: much research has been conducted on the relationship between gender and income inequality. The question is: does income inequality cause gender? Well (you scratch your head), no, of course not! Being paid less for the same work does not predict a person's gender. But if you flip the hypothesis around, gender does cause income equality. That is, being a female is likely to be a cause for the lower income a woman receives for the same work of her male counter-part. Whenever you are trying to discern the causal ordering of the two variables, think through the logic. Does it make sense for the ordering to be the other way around?

Once the causal relationship is established, we can determine if there is an association between the two variables. There are three questions we ask ourselves when looking for an association: (1) does an association exist? (2) how strong is the association? (3) what is the direction of the relationship?

The first two questions can be asked and answered about all levels of variables, but the third question, about the direction of the relationship, can only be asked of ordinal or interval-ratio variables, because direction can only be assessed in variables which are ranked or ordered. Since nominal variables are not ordered, but merely categories, we cannot discuss them in terms of direction.

To assess the first question, does an association exist, it is easiest to look at the percentages in each **conditional distribution of Y**. The conditional distribution of Y is the changes which occur in the dependent (Y) variable for each condition of the independent variable (X). If we have two categories for the independent variable of gender, then we have two conditions: male and female. The changes in the dependent variable (let's say income) are analyzed for each condition of X, first males then females. If an association exists, then there will be a change in each condition for the distribution of Y. In English, that means that we will consider an association to exist if a majority of males have high incomes and a majority of the females have a low income.

The second and third questions, how strong is the association and what is the direction of the relationship, will be discussed in the next three chapters. It is important to remember that the direction of the relationship cannot be addressed for nominal data.

MULTIPLE-CHOICE QUESTIONS

1. Bivariate means:
   a. two samples
   b. two variables
   c. two means
   d. association

2. For an association to exist, the ___________ for one variable changes for the ___________ of another variable.
   a. distribution, categories
   b. categories, distribution
   c. distribution, distribution
   d. cases, means

3. The dependent variable is thought to be ______________.
   a. the cause
   b. the effect
   c. positive
   d. X

4. The independent variable is thought to be ______________.
   a. the cause
   b. the effect
   c. positive
   d. Y

5. X is used to identify the ________ variable.
   a. independent
   b. dependent
   c. bivariate
   d. nominal

6. Y is used to identify the ________ variable.
   a. independent
   b. dependent
   c. bivariate
   d. nominal

7. A negative association is one in which the variables vary in ___________ directions.
a. the same
b. opposite
c. negative
d. positive

8. A positive association is one in which the variables vary in ___________ directions.
a. the same
b. opposite
c. negative
d. positive

9. What three questions can we ask (and answer) about a bivariate relationship?
a. Does an association exist? What is the strength? What should be the logical next step?
b. Does the independent variable cause the effect? What is the direction of the relationship? Does an association exist?
c. Does an association exist? Is there a third variable? What is the direction of the association?
d. Does an association exist? What is the strength of the association? What is the direction of the association?

10. In order to determine if an association exists, the logical first step would be to:
a. find the cumulative frequencies
b. find the percentages
c. compare the means across categories
d. include a third variable

11. Percentages are calculated in the direction of the ___________ variable.
a. independent
b. dependent
c. bivariate
d. nominal

12. When creating a bivariate table, the _____________ variable should be placed along the top of the table, and the ______________ variable should be placed along the side of the table.
a. nominal, ordinal
b. ordinal, nominal
c. dependent, independent
d. independent, dependent

13. A perfect positive relationship will have a value of _______.
   a. 0.0
   b. -1.0
   c. +1.0
   d. +10.0

14. A perfect negative relationship will have a value of _______.
   a. 0.0
   b. -1.0
   c. +1.0
   d. +10.0

15. If one variable increases and the other variable decreases, the association is said to be:
   a. positive
   b. different
   c. negative
   d. similar

WORK PROBLEMS

1. If a researcher is interested in the relationship between age and education, which variable would be independent and which would be dependent?

2. This same researcher is interested in education and age first married, hypothesizing that more educated people marry later. Which is the independent variable and which is the dependent variable?

3. Quality of life and health status are thought to be associated. Make a hypothesis, indicating which variable is independent and which is dependent.

4. Marital happiness and job satisfaction are thought to be associated. Which variable would be the independent (causal) and which would be the dependent?

5. Is there an association between these two variables?

Life Satisfaction	Job Satisfaction High	Low	
High	54	29	83
Low	31	65	96
	85	94	179

6. Is there an association between these two variables?

Marital Satisfaction	Gender Male	Female	
Very Happy	30	45	75
Moderately Happy	50	45	95
Unhappy	20	10	30
	100	100	200

7.

Attitude Toward Gun Control	Gender Male	Female	
Favor	45	42	87
Oppose	39	49	88
	84	91	175

8. A researcher is interested in knowledge of birth control between men and women. Is there an association between these two variables?

Knowledge of Birth Control	Gender Male	Female	
High	175	430	605
Moderate	512	220	732
Low	215	160	375
	902	810	1710

9. In a study of attitudes toward legalization of marijuana, the following data were collected. Are males more likely to favor legalization than females?

Legalization of Marijuana	Gender: Male	Gender: Female	
Approve	10	6	16
Disapprove	5	9	14
No Opinion	9	11	20
	24	26	50

10. Is there an association between place of residence and approval of the proposed water plant?

Approval of Water Plant	Place of Residence: Town	Place of Residence: Farm	
Yes	33	46	79
No	37	30	67
	70	76	146

DETAILED ANSWERS

1. Since age predicts when you begin school, and the older you get the more schooling you attain, then age is the independent variable. Another way to think about it is age increases the amount of schooling one has, education does not increase someone's age.

2. As stated in the hypothesis, education is the independent, or causal, variable, which is hypothesized to have a positive relationship on age first married, which is the dependent variable.

3. Hypothesis: Health status will have a positive effect on quality of life.
As one's health decreases, so too will his/her quality of life. Therefore, health status is the independent variable, and quality of life is the dependent variable.

4. The most logical hypothesis would be that job satisfaction would affect marital happiness, since if you are unsatisfied at work, where you spend most of your waking hours, you are likely to be unhappy in other parts of your life as well, which includes your relationship.

5.

Life Satisfaction	Job Satisfaction: High	Low	
High	54(64%)	29(31%)	83
Low	31(36%)	65(69%)	96
	85(100%)	94(100%)	179

Yes, there seems to be an association between these two variables. People with high life satisfaction seem to also have high job satisfaction (64%), while people with low life satisfaction seem to have low job satisfaction (69%).

6.

Marital Satisfaction	Gender: Male	Female	
Very Happy	30(30%)	45(45%)	75
Moderately Happy	50(50%)	45(45%)	95
Unhappy	20(20%)	10(10%)	30
	100(100%)	100(100%)	200

There is no strong evidence that these two variables are associated, since there are no clear distinctions between the two groups. It appears that more women state that they are very happy with their marriage (45% compared to 30% for males), yet the moderately happy category is almost even across both groups (50% for males, 45% for females).

7.

AttitudeToward Gun Control	Gender: Male	Female	
Favor	45(54%)	42(46%)	87
Oppose	39(46%)	49(54%)	88
	84(100%)	91(100%)	175

Yes, there appears to be an association between these two variables, although it seems to be weak.

8.

Knowledge of Birth Control	Gender: Male	Female	
High	175(19%)	430(53%)	605
Moderate	512(57%)	220(27%)	732
Low	215(24%)	160(20%)	375
	902(100%)	810(100%)	1710

These two variables appear to be associated, since, when comparing the conditional distribution of Y, women appear to be more knowledgeable than men.

9.

Legalization of Marijuana	Gender: Male	Gender: Female	
Approve	10(42%)	6(23%)	16
Disapprove	5(21%)	9(35%)	14
No Opinion	9(37%)	11(42%)	20
	24(100%)	26(100%)	50

Since there appears to be a weak association between the variables, we could conclude that males are more likely to favor legalization than females.

10.

Approval of Water Plant	Place of Residence: Town	Place of Residence: Farm	
Yes	33(47%)	46(61%)	79
No	37(53%)	30(39%)	67
	70(100%)	76(100%)	146

Farm residents appear more likely to approve the water plant than town residents (61% vs. 53%). There seems to be an association between these two variables.

CHAPTER FOURTEEN
ASSOCIATION BETWEEN VARIABLES MEASURED at the NOMINAL LEVEL

Three measures of association for nominal data are introduced in chapter 14 of your text, phi (Φ) , Cramer's V, and lambda (λ). The first two measures, phi and V, are chi square based measures, which means they are calculated using the chi square statistic. Phi is used for 2x2 tables, while V is used for tables larger than 2x2. Both are good measures, but interpretation is a bit sticky. Because there is no real interpretation, we can only make generalizations. Thus, a phi of 0.33 seems a bit weak, and we know it is weaker than a phi of 0.54, but exactly how much weaker is unknown. Therefore it might be better to use another type of measure, thus we use lambda.

Lambda is a PRE measure, which stands for Proportional Reduction in Error, and indicates the amount of error we reduce in predicting the dependent variable with the knowledge of the independent variable. In your textbook, Healey introduces a nice example of the logic (page 330). Basically it works like this: let's say you have to predict the height of every person walking through a door as either tall (over 5'9") or short (under 5'9"), with no information about the person. Well, you have a 50/50 chance of correctly guessing their height. Now, what if you are told that the next person to walk through that door will be a female. We logically can say that women, on average, are shorter than 5'9", therefore, if we predict that she, too, will be short, there is a good possibility that we will be correct. Therefore, knowledge of the person's gender will likely reduce the number of errors we make in predicting height.

Try to keep this example in mind whenever you are working with PRE measures (which will be throughout the rest of the text). Remember that you are trying to reduce the number of errors you are making in predicting the dependent variable **with knowledge of** the independent variable.

Lambda then can be interpreted as a percentage. If we have a lambda of 0.55, we can convert that to a percentage and state that we have reduced the amount of error in predicting the dependent variable by 55% with knowledge of the independent variable. PRE measures range from zero to 1.0, with zero indicating no improvement whatsoever in our prediction of the dependent variable with the independent variable, and 1.0 indicating that the independent variable perfectly predicts the dependent variable.

One problem with lambda that you might encounter is that when one row marginal is exceptionally large, lambda might be zero. This does not necessarily mean that there is no association between these two variables, only that the large row marginal has made lambda act strangely. If this is the case, use a chi square based measure of association (either phi or V) to assess the strength of the relationship.

MULTIPLE-CHOICE QUESTIONS

1. The measures introduced in this chapter are used on __________ level(s) of measurement variables.
   a. nominal
   b. ordinal
   c. interval-ration
   d. all

2. The chi-square based measures of association are ________ and ________.
   a. lambda, phi
   b. lambda, Cramer's V
   c. phi and V
   d. phi and chi

3. The phi measure is used on ______ tables.
   a. ordinal
   b. 2x2
   c. larger than 2x2
   d. multivariate

4. The Cramer's V measure is used on ______ tables.
   a. ordinal
   b. 2x2
   c. larger than 2x2
   d. multivariate

5. PRE is an acronym for:
   a. Percentile Reduced Error
   b. PRoportional Error
   c. Purposeful Reducing Error
   d. Proportional Reduction of Error

6. The PRE measure of association for nominal data is ___________.
   a. phi
   b. Cramer's V
   c. lambda
   d. chi square

7. PRE means that we reduce the error in the __________ with knowledge of the __________.
   a. PRE measure, independent variable
   b. independent variable, dependent variable
   c. PRE measure, chi square
   d. dependent variable, independent variable

8. When computing lambda, we need to find ________ and ________.
   a. E_1, E_2
   b. chi square, degrees of freedom
   c. degrees of freedom, f ratio
   d. phi, V

9. The symbol ____ is used to designate lambda.
   a. α
   b. Γ
   c. Φ
   d. λ

10. The symbol ____ is used to designate phi.
   a. α
   b. Γ
   c. Φ
   d. λ

11. The independent variable is the ________ variable.
   a. cause
   b. effect
   c. unknown
   d. known

12. The dependent variable is the ________ variable.
   a. cause
   b. effect
   c. unknown
   d. known

13. The independent variable is designated with ________ symbol.
   a. Φ
   b. X
   c. λ
   d. Y

14. The dependent variable is designated with ________ symbol.
   a. Φ
   b. X
   c. λ
   d. Y

15. The dependent variable is displayed along the _________ of the table, while the independent variable is displayed along the ________ .
   a. top, side
   b. side, bottom
   c. top, bottom
   d. side, top

WORK PROBLEMS

For each of the problems below, chose the **best** measure of association and write up your conclusions.

1.

Life Satisfaction	Job Satisfaction: High	Job Satisfaction: Low	
High	54	29	83
Low	31	65	96
	85	94	179

χ^2=19.2

2.

Marital Satisfaction	Gender: Male	Gender: Female	
Very Happy	30	45	75
Moderately Happy	50	45	95
Unhappy	20	10	30
	100	100	200

χ^2=6.66

3.

Attitude Toward Gun Control	Gender Male	Female	
Favor	45	42	87
Oppose	39	49	88
	84	91	175

χ^2=0.96

4.

Knowledge of Birth Control	Gender		
	Male	Female	
High	175	430	605
Moderate	512	220	732
Low	215	160	375
	902	810	1710

χ^2=227.7

5.

Amount of Housework	Gender		
	Male	Female	
All	15	46	61
Most	20	53	73
Some	69	27	96
None	11	9	20
	115	135	250

χ^2=7.815

6.

Victim of Crime	Area of Residence		
	Urban	Suburban	
Yes	40	25	65
No	25	35	60
	65	60	125

χ^2=4.93

7.

Approval of President	Political Party Preference: Republican	Democrat	
Yes	14	29	43
No	24	11	35
	38	40	78

$\chi^2=9.87$

8.

Political Party Affiliation	Liberal	Conservative	
Republican	3	25	28
Democrat	29	2	31
Independent	10	8	18
	42	35	77

$\chi^2=40.9$

9.

Legalization of Marijuana	Gender: Male	Female	
Approve	10	6	16
Disapprove	5	9	14
No Opinion	9	11	20
	24	26	50

$\chi^2=2.23$

10.

Grades on Stats Exam	Gender: Male	Female	
High	25	15	40
Low	10	10	20
	35	25	60

$\chi^2=0.89$

DETAILED ANSWERS

1.

Life Satisfaction	Job Satisfaction High	Low	
High	54	29	83
Low	31	65	96
	85	94	179

χ^2=19.2

E1=83
E2=31+29=60

$$\lambda = \frac{E1 - E2}{E1} = \frac{83 - 60}{83} = 0.28$$

Therefore, we reduce the errors in predicting life satisfaction by 28% with knowledge of job satisfaction.

2.

Marital Satisfaction	Gender Male	Female	
Very Happy	30	45	75
Moderately Happy	50	45	95
Unhappy	20	10	30
	100	100	200

χ^2=6.66

E1= 200 - 95 = 105
E2= 50 + 55 = 105

$$\lambda = \frac{E1 - E2}{E1} = \frac{105 - 105}{105} = 0$$

$$V=\sqrt{\frac{X^2}{N(r-1)}}=\sqrt{\frac{6.66}{(200)(1)}}=\sqrt{0.33}=0.18$$

Therefore, there is a very weak association between these two variables. We cannot talk about the PRE measure since lambda was equal to zero.

3.

Attitude Toward Gun Control	Gender Male	Female	
Favor	45	42	87
Oppose	39	49	88
	84	91	175

$\chi^2=0.96$

E1 = 87
E2 = 39 + 42 = 81

$$\lambda = \frac{E1 - E2}{E1} = \frac{87 - 81}{87} = 0.07$$

Therefore, we reduce the amount of error in predicting attitudes toward gun control by 7% with knowledge of gender. There is a very weak association and a non-significant relationship ($\chi^2=0.96$) between these two variables.

4.

Knowledge of Birth Control	Gender Male	Female	
High	175	430	605
Moderate	512	220	732
Low	215	160	375
	902	810	1710

$\chi^2=227.7$

E1 = 980
E2 = 390 + 380 = 770

$$\lambda = \frac{E1 - E2}{E1} = \frac{980 - 770}{980} = 0.21$$

Therefore, we make 21% fewer errors in predicting knowledge of birth control by including the variable gender.

5.

Amount of Housework	Gender: Male	Female	
All	15	46	61
Most	20	53	73
Some	69	27	96
None	11	9	20
	115	135	250

χ^2=7.815
E1= 250 - 96 = 154
E2= (115-69)+ (135-53) = 46 + 82 = 128

$$\lambda = \frac{E1 - E2}{E1} = \frac{154 - 128}{154} = 0.17$$

Therefore we will make 17% fewer errors in our prediction of the amount of housework a person will do by first knowing his/her gender.

6.

Victim of Crime	Area of Residence: Urban	Suburban	
Yes	40	25	65
No	25	35	60
	65	60	125

χ^2=4.93

E1= 60
E2 = 25 + 25 = 50

$$\lambda = \frac{E1 - E2}{E1} = \frac{60 - 50}{60} = 0.17$$

We have reduced the amount of error in predicting whether or not a person has been a victim of crime with knowledge of area of residence.

7.

Approval of President	Political Party Preference: Republican	Democrat	
Yes	14	29	43
No	24	11	35
	38	40	78

χ^2=9.87

E1 = 35
E2 = 14 + 11 = 25

$$\lambda = \frac{E1 - E2}{E1} = \frac{35 - 25}{35} = 0.29$$

We have reduced the amount of error in predicting presidential approval by 29% with knowledge of the independent variable, political party preference.

8.

Political Party Affiliation	Liberal	Conservative	
Republican	3	25	28
Democrat	29	2	31
Independent	10	8	18
	42	35	77

χ^2=40.9

E1 = 77 - 31 = 46
E2 = (42-29) + (35-25) = 13 + 10 = 23

$$\lambda = \frac{E1 - E2}{E1} = \frac{46 - 23}{46} = 0.50$$

We have reduced the amount of errors in prediction political party affiliation with knowledge of political beliefs.

9.

Legalization of Marijuana	Gender: Male	Female	
Approve	10	6	16
Disapprove	5	9	14
No Opinion	9	11	20
	24	26	50

χ^2=2.23

E1 = 50 - 20 = 30
E2 = (24 -10) + (26 -11) = 14+15 = 29

$$\lambda = \frac{E1 - E2}{E1} = \frac{30 - 29}{30} = 0.03$$

We reduce our error only slightly (3%) in predicting attitudes toward legalization of marijuana with knowledge of gender.

10.

Grades on Stats Exam	Gender: Male	Female	
High	25	15	40
Low	10	10	20
	35	25	60

$\chi^2 = 0.89$

E1 = 20
E2 = 20

$$\lambda = \frac{E1 - E2}{E1} = \frac{20 - 20}{20} = 0$$

$$\phi = \sqrt{\frac{\chi^2}{N}} = \sqrt{\frac{0.89}{(60)}} = \sqrt{0.015} = 0.12$$

Lambda indicates that there is no association between these two variables. With the non-significant chi square, we probably could have stopped there. But just to be sure, we computed the phi measure as well, which also indicates that there is a very weak relationship (if any at all).

CHAPTER FIFTEEN
ASSOCIATION BETWEEN VARIABLES MEASURED AT THE ORDINAL LEVEL

When we have two ordinal level variables, we can ask the questions: (1) is there an association?, (2) how strong is the association?, and (3) what direction is the association? Unlike nominal data, ordinal data is ranked and ordered, therefore we can refer to the relationship between two ordinal variables in terms of direction: Is the relationship positive or negative? A positive relationship, again, is one in which both variables vary in the same direction; either they both increase together, or they both decrease together. A negative relationship, then, is one in which the variables vary in opposite directions; as one increases, the other decreases.

In this chapter, you are introduced to measures of association for ordinal variables. Two important, yet distinct, measures under this heading are gamma and Spearman's rho.

Gamma (G) is used when you have two ordinal variables (or one ordinal and one interval-ratio) in a bivariate table of only a few categories (less than six categories per variable). This type of data, also called **collapsed ordinal data**, is used quite a lot in social science research, such as in scales which range from Strongly Agree to Strongly Disagree, or in social class categories, such as upper, middle, working, and lower.

Spearman's rho (r_s), on the other hand, is used when you have **continuous ordinal data**, or data that looks like interval-ratio data. For example, scales with a wide range of possibilities, such as the prestige scales which range from 10-89 are not true interval-ratio variables, yet it would be difficult to break this scale into five or six categories.

Gamma, which is used for ordinal data in a bivariate table, is a funny statistic to grasp, but bear with us and we'll get you through it! Basically it works like this: remember in PRE measures for nominal data, we were trying to predict one variable with the knowledge of another variable (think back to the example of predicting a person's height with knowledge of the independent variable). In gamma, you are still trying to predict the dependent variable with knowledge of the independent variable, but now you are trying to do so with **pairs of cases** (sounds strange, doesn't it!).

O.k., let's try this: Ruthy and Cindy are both on a swim team. If we want to predict who is in the faster swim lane (the six lanes are ordered from fastest to slowest), with no other information we have a one-in-three chance of correctly guessing (they are either in the same lane, Ruthy is faster, or Cindy is faster). But let's say we know how often they work out: Cindy works out a lot, Ruthy works out fairly often. Given this information, we know that Cindy is higher on the independent variable (amount of workouts) than Ruthy, therefore, we can predict that she will also be higher on the dependent variable (the swim lane). By utilizing **pairs** of cases (Ruthy and Cindy) on both variables, we can better predict the outcome of the dependent variable. Gamma utilizes all pairs of cases ranked the same (N_s) and all pairs of cases ranked differently (N_d) to determine the strength of the association. Gamma is also a PRE measure, therefore, we can refer to reducing the amount of errors in predicting the dependent variable with knowledge of the independent variable.

A very important point to remember when calculating gamma is to look at the way the data are set up in the bivariate table. Be sure that the low-lows are together in one corner of the table and begin your calculations from there.

Spearman's rho, on the other hand, uses continuous ordinal data. This data will often look like interval-ratio data, but it's not. There usually is no true zero-point, and the distance between two scores is not necessarily known. Let's say we have a group of swimmers in which we know how often they work out and want to know where they placed in a recent meet. Here's how the data looked:

SWIMMERS	WORKOUTS/WEEK	RACE FINISH
Carl	7	1
Cindy	6	4
Diane	9	3
Marlene	6	5
Ruthy	4	6
Steve	12	2
Tom	3	7

We can rank how often each of the swimmers works out, as well as where they finished in the race. Note, though, that the independent variable (workouts per week) is interval-ratio. Because the other variable is ordinal, though, we must use an ordinal measure of association. **Always use the measure appropriate for the lowest level of measurement between two variables!**

We can see from this data that Steve works out the most, therefore we assign him the rank of <u>one</u> on the *first* variable. Carl came in first in the race, so he gets the rank of <u>one</u> on the *second* variable. We continue on in this manner until every person receives a rank on the first variable and a rank on the second variable. These ranks are then compared and the rho statistic is computed. Spearman's rho can range from -1 to +1 indicating the strength of the relationship. Like gamma, a -1 indicates a perfect negative relationship, while a +1 indicates a perfect positive relationship. Unfortunately, though, rho by itself cannot be interpreted as a PRE measure. Instead you must square rho (r_s^2) for it to be interpreted as such.

The rho for the above data is +0.88 and r_s^2=0.77. Therefore, we would conclude that there is a fairly strong, positive association between these two variables. We reduce the amount of errors in predicting the finishing place of the swimmers in the swim meet by 77% with knowledge of the number of times they work out per week.

A final important step to remember in measures of associations is that these measures are simply **descriptions** about our **sample**. If we want to infer our sample results to the rest of the population (which, as I stated earlier, is what we <u>live</u> for) we need to compute either a Z-test (for gamma) or a t-test (for Spearman's rho). These tests will indicate whether our sample can be inferred to the rest of the population (it is significant) or not.

The null hypothesis for both of these tests is that there will be no association between these two variables in the population. If we fail to reject the null hypothesis, then we must conclude that the association we found between the two variables in our sample is not likely to be found in the population, and occurs by chance alone. If we reject the null hypothesis, then we conclude that there is a high probability that this relationship exists in the populations from which the sample was drawn.

MULTIPLE-CHOICE QUESTIONS

1. Gamma is an appropriate measure for __________ level of measurement variables.
   a. nominal
   b. ordinal
   c. interval-ratio
   d. all of the above

2. Gamma uses ________ to determine association.
   a. means
   b. populations
   c. row marginals
   d. pairs of cases

3. Somer's d utilizes ____________ to determine association.
   a. cases tied on the dependent variable
   b. column marginals
   c. row marginals
   d. cases tied on the independent variable

4. Gamma is appropriate for __________ data, whereas Spearman's rho is appropriate for _________ data.
   a. ordinal, interval-ratio
   b. collapsed, continuous
   c. bivariate, multivariate
   d. ordinal, nominal

5. The term N_s is the number of pairs of cases ranked ____________ on the two variables.
   a. simultaneously
   b. systematically
   c. the same
   d. symbolically

6. The term N_d is the number of pairs of cases ranked ____________ on the two variables.
   a. depressionistically
   b. diagonally
   c. directionally
   d. differently

7. The term T_y is the number of pairs of cases tied on the ____________________.
a. independent variable
b. dependent variable
c. column marginals
d. row marginals

8. The term T_x is the number of pairs of cases tied on the ____________________.
a. independent variable
b. dependent variable
c. column marginals
d. row marginals

9. The null hypothesis for gamma is:
a. $\mu_1=\mu_2$
b. $\rho=0.00$
c. $\gamma=0.00$
d. $\rho_1=\rho_2$

10. The null hypothesis for Spearman's rho is:
a. $\mu_1=\mu_2$
b. $\rho=0.00$
c. $\gamma=0.00$
d. $\rho_1=\rho_2$

11. Significance tests are conducted for measures of association to determine if:
a. the two variables are associated in the sample
b. there is a significant difference between the sample and the population
c. the two variables are associated in the population
d. the means are significantly different

12. The ______ distribution is used for gamma, while the ______ distribution is used for Spearman's rho.
a. chi square, f
b. t, Z
c. Z, t
d. Z, chi square

13. Spearman's rho uses __________ to determine if two variables are associated.
a. means
b. standard deviations
c. ranks on both variables
d. collapsed data

14. The sum of D in Spearman's rho will always equal ____________.
a. 0
b. 1
c. the mean
d. Z

15. If the two variables are measured at different levels, you should use the __________ measure of association.
a. higher
b. lower
c. ordinal
d. tau-b

WORK PROBLEMS

1. A researcher wanted to know if noise tolerance was associated with tolerance for children. She collected the following data. How strong is the association?

TOLERANCE FOR CHILD.	NOISE TOLERANCE LOW	MED.	HIGH	
LOW	31	5	6	42
MED	14	23	15	52
HIGH	7	15	17	39
	52	43	38	133

2. Is intelligence related to how well a person will do in statistics class?

SUCCESS IN STATS CLASS	INTELLIGENCE LOW	MED	HIGH	
LOW	10	7	8	25
MED	11	12	13	36
HIGH	9	11	9	29
	30	30	30	90

3. Media campaigns were used in several developing countries to promote family planning. Researchers were brought in to find out if these campaigns were a success. Is media consumption associated with knowledge of family planning methods?

KNOWL. OF FAMILY PLANNING	MEDIA CONSUMPTION LOW	HIGH	
LOW	262	71	333
MODERATE	74	197	271
HIGH	43	275	318
	379	543	922

4. Is the popularity of an older sibling associated with the popularity of the younger sibling? Sixty-two younger siblings were asked to rate the popularity of their older sibling and their own popularity. How strong is the association?

POPULARITY OF YOUNGER SIBLING	POPULARITY OF OLDER SIBLING LOW	HIGH	
LOW	21	10	31
HIGH	14	17	31
	35	27	62

5. Is class ranking associated with popularity ranking?

Respondent	Class Ranking	Popularity Ranking
1	1	10
2	7	5
3	2	2
4	5	6
5	11	7
6	4	12
7	10	9
8	3	1
9	6	4
10	8	8
11	12	11
12	9	3

6. Is income associated with attitudes toward welfare?

ATTITUDE TOW. WELFARE	INCOME LOW	MODERATE	HIGH	
STRONGLY OPPOSE	58	57	68	183
OPPOSE	69	71	56	196
NO OPINION	60	121	26	207
FAVOR	75	149	40	264
STRONGLY FAVOR	63	52	35	150
	325	450	225	1000

7. Can the results found in Question 6 above be inferred to the population? Write a sentence or two discussing the results.

8. Much research has been conducted in the field of stratification. Specifically, the question has been asked: Does father's prestige lead to son's prestige? Using the data below, how strongly associated are these two variables?

Respondent	FATHER'S PRESTIGE	SON'S PRESTIGE
1	50	62
2	25	31
3	87	56
4	33	47
5	46	51
6	86	86
7	52	57
8	48	50
9	71	67
10	86	89
11	75	32
12	23	74
13	30	34
14	18	46
15	80	72

9. Is the Spearman's rho found in Question 8 above significant? What does the significance indicate?

10. Is the amount of effort you put into your schoolwork associated with grades?

	EFFORT IN SCHOOLWORK		
GRADES	LOW	HIGH	
LOW	37	32	69
HIGH	7	14	21
	44	46	90

11. It has been said that people rarely marry above "their station." Below is some data regarding the SES of respondent's family (their "station") and the SES of their spouse's family. Is this statement true?

SES OF SPOUSE'S	SES OF FAMILY				
FAMILY	UPPER	MIDDLE	WORKING	LOWER	
UPPER	221	71	21	4	317
MIDDLE	20	560	370	33	983
WORKING	7	203	420	100	730
LOWER	2	166	189	113	470
	250	1000	1000	250	2500

12. The number of friends a person has is hypothesized to have a positive effect on his/her quality of life. Given the data below, how strong is the relationship between these two variables?

Respondent	NO. OF FRIENDS	QUALITY OF LIFE
JOHN	5	7
MARY	9	10
JANE	7	6
SUSAN	3	5
GEORGE	1	3
JEFF	10	10
CLAIRE	4	5
STUART	15	8
SHERYL	6	9
EMILY	3	6
BILL	5	7

DETAILED ANSWERS

1.

$$N_s = \begin{array}{ll} 31(23+15+15+17) & = 2170 \\ +14(15+17) & + 448 \\ + 5(15+17) & + 160 \\ +23(17) & + \underline{391} \\ & 3169 \end{array}$$

$$N_d = \begin{array}{ll} 6(23+15+14+7) & = 354 \\ +15(15+7) & + 330 \\ + 5(14+7) & + 105 \\ +23(7) & + \underline{161} \\ & 950 \end{array}$$

$$G = \frac{N_s - N_d}{N_s + N_d} = \frac{3169-950}{3160+950} = \frac{2219}{4119} = 0.54$$

Therefore, we reduce our errors by 54% when predicting tolerance of children by first knowing tolerance of noise.

2.

$$N_s = \begin{array}{ll} 10(12+11+13+ 9) & = 450 \\ +11(11+ 9) & + 220 \\ + 7(13+ 9) & + 154 \\ +12(9) & + \underline{108} \\ & 932 \end{array}$$

$$N_d = \begin{array}{ll} 8(12+11+11+9) & = 344 \\ +13(11+ 9) & + 260 \\ + 7(11+ 9) & + 140 \\ +12(9) & + \underline{108} \\ & 852 \end{array}$$

$$G = \frac{N_s - N_d}{N_s + N_d} = \frac{932-852}{932+852} = \frac{80}{84} = 0.04$$

Therefore, intelligence has very little to do with how well one will do in statistics course. In fact, only four percent of errors are reduced in predicting class success with the variable intelligence.

3.

$$
\begin{aligned}
N_s &= 262(197+275) & 123664 \\
+ & \quad 74(275) = & \underline{20350} \\
& & 144014
\end{aligned}
$$

$$
\begin{aligned}
N_d &= 71(74+43) = & 8307 \\
& +197(43) = & \underline{8471} \\
& & 16778
\end{aligned}
$$

$$G = \frac{N_s - N_d}{N_s + N_d} = \frac{144014-16778}{144014+16778} = \frac{127236}{160792} = 0.79$$

Yes, media consumption is associated with knowledge of family planning. We can reduce our error in predicting family planning knowledge by first knowing the amount of media consumed by 79%.

4.

$N_s = 21(17) = \quad 357$

$N_d = 10(14) = \quad 140$

$$G = \frac{N_s - N_d}{N_s + N_d} = \frac{357-140}{357+140} = \frac{217}{497} = 0.44$$

There seems to be a moderately strong, positive association between these two variables. We reduce our errors of prediction by 44% with knowledge of the popularity of the older sibling.

5.

Respondent	Ranking	Class Popularity Ranking	D	D^2
1	1	10	-9	81
2	7	5	2	4
3	2	2	0	0
4	5	6	-1	1
5	11	7	4	16
6	4	12	-8	64
7	10	9	1	1
8	3	1	2	4
9	6	4	2	4
10	8	8	0	0
11	12	11	1	1
12	9	3	6	36
			$\sum D = 0$	$\sum D^2 = 212$

$$r_s = 1 - \frac{6\sum D^2}{N(N^2-1)} = 1 - \frac{6(212)}{12(143)} = 1 - 0.76 = 0.26$$

Therefore, there is a weak, but positive association between class ranking and popularity. We reduce our errors by only 7% ($r_s^2=0.07$) in predicting popularity ranking by first knowing class ranking.

6.

$$
\begin{aligned}
N_s = \;& 58(71+121+149+56+52+26+40+35) && = 31900 \\
& +69(121+149+52+26+40+35) && + 29187 \\
& +60(149+52+26+40+35) && + 18120 \\
& +75(52+35) && + 6525 \\
& +57(56+26+40+35) && + 8949 \\
& +71(26+40+35) && + 7171 \\
& +121(40+35) && + 9075 \\
& +149(35) && + \underline{5215} \\
& && = 116142
\end{aligned}
$$

$$
\begin{aligned}
N_d = \;& 68(71+121+149+52+69+60+75+63) && = 44880 \\
& +56(121+149+52+60+75+63) && + 29120 \\
& +26(149+52+75+63) && + 8814 \\
& +40(52+63) && + 4600 \\
& +57(69+60+75+63) && + 15219 \\
& +71(60+75+63) && + 14058 \\
& +121(75+63) && + 16698 \\
& +149(63) && + \underline{9387} \\
& && = 142776
\end{aligned}
$$

$$G = \frac{N_s - N_d}{N_s + N_d} = \frac{116142-142776}{116142+142776} = \frac{-26634}{258918} = -0.103$$

There is a weak, negative association between income and attitudes toward welfare. As income increases, attitudes toward welfare decreases. We reduce our error in predicting attitude about welfare by 10% with knowledge of income.

7.

$$Z = G\sqrt{\frac{N_s+N_d}{N(1-G^2)}} = 0.17\sqrt{\frac{102259+142776}{1000(1-0.17^2)}} = 0.17\sqrt{252.61} = 2.70$$

Z(obtained) +2.70 falls in the critical region ($\pm$1.96). Therefore we reject the null hypothesis of no association. We are 95% confident that the variables are related in the populations from which they were drawn.

8.

FATH PREST	RANK	SON'S PREST	RANK	D	D^2
50	8	62	6	2	4
25	13	31	15	-2	4
87	1	56	8	-7	49
33	11	47	11	0	0
46	10	51	9	1	1
86	2.5	86	2	0.5	0.25
52	7	57	7	0	0
48	9	50	10	-1	1
71	6	67	5	1	1
86	2.5	89	1	1.5	2.25
75	5	32	14	-9	81
23	14	74	3	11	121
30	12	34	13	-1	1
18	15	46	12	3	9
80	4	72	4	0	0
				$\sum D=0$	$\sum D^2=274.5$

$$r_s=1-\frac{6\sum D^2}{N(N^2-1)}=1-\frac{6(274.5)}{15(224)}=1-0.49=0.51$$

There is a moderately strong, positive relationship between these two variables. We will reduce our errors by 26% ($r_s^2=0.26$) when predicting son's prestige given father's prestige.

9.

$$t=\sqrt{\frac{N-2}{1-r_s^2}}=\sqrt{\frac{13}{1-0.26}}=\sqrt{17.57}=4.19$$

T(obtained) is 4.19; t(critical) with two degrees of freedom at the 0.05 alpha level is 4.303. Therefore we fail to reject the null hypothesis of no association. We cannot state with any confidence that our findings can be inferred to the rest of the population.

10. N_s= 37(14) = 518

N_d= 32(7) = 224

$$G = \frac{N_s - N_d}{N_s + N_d} = \frac{518-224}{518+224} = \frac{294}{742} = 0.40$$

There is a moderately strong, positive association between these two variables.

11.

N_s=	221(560+203+166+370+420+189+33+100+113)	=	476034
	20(203+166+420+189+100+113)	+	23820
	7(166+189+113)	+	3276
	71(370+420+189+33+100+113)	+	86975
	560(420+189+100+113)	+	460320
	203(189+113)	+	61306
	21(33+100+113)	+	5166
	370(100+113)	+	78810
	420(113)	+	47460
		=	1243167

N_d=	4(370+420+189+560+203+166+20+7+2)	=	7748
	33(420+89+203+166+7+2)	+	29040
	100(189+166+2)	+	35700
	21(560+203+166+20+7+2)	+	20118
	370(203+166+7+2)	+	139860
	420(166+2)	+	70560
	71(20+7+2)	+	2059
	560(7+2)	+	5040
	203(2)	+	406
		=	310531

$$G = \frac{N_s - N_d}{N_s + N_d} = \frac{1243167-310531}{1243167+310531} = \frac{932636}{1553698} = 0.60$$

Therefore, there is a moderately strong, positive relationship between these two variables. We reduce our error by 60% in predicting the SES of the family a person will marry into by first knowing the SES of the person's biological family.

12.

NO. OF FRIENDS	RANK	QUAL. LIFE	RANK	D	D^2
5	6.5	7	5.5	1	1
9	3	10	1.5	1.5	2.25
7	4	6	7.5	-3.5	12.25
3	9.5	5	9.5	0	0
1	11	3	11	0	0
10	2	10	1.5	0.5	0.5

4	8	5	9.5	-1.5	1.5
15	1	8	4	- 3	3
6	5	9	3	2	2
3	9.5	6	7.5	2	2
5	6.5	7	5.5	1	1
				$\sum D = 0$	$\sum D^2 = 36$

$$r_s = 1 - \frac{6\Sigma D^2}{N(N^2-1)} = 1 - \frac{6(36)}{11(120)} = 1 - 0.16 = 0.84$$

Therefore there is a fairly strong, positive relationship between these two variables. When we square this number, we find that we make 71% fewer errors in predicting quality of life with knowledge of number of friends.

CHAPTER SIXTEEN
ASSOCIATION BETWEEN VARIABLES MEASURED AT THE INTERVAL-RATIO LEVEL

When we are interested in the association between two interval-ratio variables, we want to keep in mind the three questions: (1) does an association exist?, (2) how strong is the association?, and (3) what is the direction of the association?

In order to answer the first question, does an association exist for nominal and ordinal data we used a bivariate table. For interval-ratio variables, we use what is called a **scattergram**. A scattergram is a graph that displays the values of the independent variable along the horizontal axis, and the values of the dependent variable along the vertical axis. The values of each respondent on both variables are then plotted on the graph. The display of the plotted values then gives us an idea of whether or not these variables are associated.

If the values of the dependent variable increase with the value of the independent variable, then we can say that there is a positive association. If the values of the dependent variable decrease with the value of the independent variable, then we can say that there is a negative association. If the distribution of the plot is basically flat or there is no pattern to the plot, then we conclude there is no association between these two variables.

The second step in analyzing the association between these two variables is to find the best-fitting regression line. The regression line is the line that best fits the points on the scattergram, and indicates the strength of the relationship. If the line is steep, then we can conclude that there is a strong association between the two variables. If the line is flat, then we conclude that there is a very weak or no association between the two variables. The equation for the regression line is:

$$Y = a + bX$$

where a = the Y-intercept
b = the slope of the line
and X = a score on the independent variable

The Y-intercept is where the regression line crosses the Y axis; the slope is the amount of change in Y for every one-unit change in X; and X is the independent variable which you are using to predict the value on the dependent variable.

Basically it works like this: remember when we speak in terms of PREs we are reducing the amount of error in <u>predicting</u> the dependent variable with knowledge of the independent variable. With regression, we are attempting to <u>predict</u> the dependent variable if the independent variable is some given amount. Regression, though, is <u>not</u> a PRE. In order to analyze the relationship in terms of the PRE measure we must use the **coefficient of determination (r^2)**. But first we must find r.

Pearson's r, or correlation, is a measure of the strength of the association between the two variables. It is a widely used measure, but cannot be interpreted as a PRE. Instead, squaring the value makes it possible to interpret in this manner. When interpreting r^2, we refer to the amount of variation <u>explained</u> in the dependent variable by the independent variable.

Thus, if we have a correlation (r) of -0.50, we say that we have a moderately strong, negative association and that 25% (0.50^2) of the variation in the dependent variable can be explained by the independent variable.

Finally, significant tests should also be calculated to insure that our sample findings can be inferred to the population. Using the t-distribution, the null hypothesis will be that there is no association between the two variables in the population. If the null hypothesis is rejected, then we can state with some certainty that our results are not by chance alone, but do exist in the population.

MULTIPLE-CHOICE QUESTIONS

1. In order to determine if an association exists between two interval-ratio variables, a ________________ should be constructed.
   a. bivariate table
   b. distribution table
   c. scattergram
   d. regression equation

2. To assess the strength of the relationship, a ____________ should be used.
   a. scattergram
   b. bivariate table
   c. gamma
   d. Pearson's r

3. The PRE measure for interval-ratio variables is:
   a. r
   b. r^2
   c. b
   d. Y

4. The term "conditional means of Y" refers to:
   a. the mean of all scores on Y for each value of X
   b. the mean of all scores on X for each value of Y
   c. the mean of each cell frequency across the Y categories.
   d. the mean of each cell frequency across each value of X

5. A linear relationship is one in which the relationship between two variables approximates:
   a. a downward trend
   b. an arc
   c. a positive association
   d. a straight line

6. In the regression equation, the term "a" is used to denote the:
   a. slope
   b. total variation explained
   c. Y-intercept
   d. regression line

7. In the regression equation, the term "b" is used to denote the:
   a. slope
   b. total variation explained
   c. Y-intercept
   d. regression line

8. In the regression equation, the term "X" refers to:
   a. the amount of variance unexplained
   b. the total variation explained
   c. a score on the independent variable
   d. a predicted score on the dependent variable

9. Homoscedasticity refers to the assumption that:
   a. the variance of the sample is equal to the variance of the population
   b. the bivariate relationship is linear
   c. the variance of the Y scores is uniform across all values of X
   d. the variables are measured on the same scale

10. Unexplained variation refers to:
   a. the proportion of variation in Y not accounted for by X
   b. the proportion of variation in X not accounted for by Y
   c. the amount of variation explained by both the independent and dependent variables
   d. the spread of the Y scores around the mean

11. Unexplained variation refers to:
a. the proportion of variation in Y not accounted for by X
b. the proportion of variation in X not accounted for by Y
c. the amount of variation explained by both the independent and dependent variables
d. the spread of the Y scores around the mean

12. Significant tests are conducted to determine if:
a. linearity exists in the population
b. the relationship found in the sample can be inferred to the population
c. the assumption of homoscedasticity has been met
d. the unexplained variance is linear in shape to the explained variance

13. In determining significance for Pearson's r, the ________ distribution is used.
a. Z
b. t
c. F
d. chi square

14. The term Y' is used to denote ____________________.
a. the predicted score on X
b. the predicted score on Y
c. the amount of variation explained in Y
d. the Y-intercept

15. When describing the PRE measure, we talk about the amount of ______________ explained by the independent variable.
a. linearity
b. variation
c. prediction
d. homoscedasticity

WORK PROBLEMS

1. After an exam, a professor asked each student to disclose the number of hours she/he studied for the exam. Each student's hours of study and grade are presented below. Is there an association?

Hours of Study	Exam Grade
5	64
1	52
6	76
3	71
4	74
9	81
11	80
14	83
2	69
1	56
7	79
4	93
6	91
3	85
8	88
12	96

2. Contemplating graduate school, Matt decided to conduct a survey on income and years of education. Specifically he was interested in the amount of increase in income for every additional year of education. Should he go to grad school?
(Hint: For ease of calculations, income has been divided by 1000, so the value of 24 is really $24,0000, and so on)

Scores		Scores (cont)	
Education	Income	Education	Income
12	24	8	12
10	14.5	23	11
16	41.7	14	29
16	32.5	15	27
18	47	12	25
21	157	13	24
22	32.5	11	14
14	28	12	17.4
16	34	16	87
18	50	19	98

3. A researcher is interested in the effects of income on the number of children a couple decides to have. Is there an association between the two variables? (Hint: For ease of calculations, income has been divided by 1000, so the value of 24 is really $24,0000, and so on)

Income	Number of Children
18	4
55	3
47	2
35	3
12	5
28	2
108	3
36	3
42	2
155	1
31	3
45	2
10	4
129	1
22	4
9	8

4. A sociologist wondered if education led to the postponement of marriage. From the data below, what can you conclude?

Education	Age Married
10	16
12	18
15	21
16	20
12	22
16	29
18	24
14	22
12	25
18	21
16	24
12	19
11	17
14	20

5. Is there an association between number of hours worked per week and income? Income measured in 1/1000.

Hrs Work/week	Income
40	47
75	157
42	32.5
34	28
40	34
42	50
26	12
40	11
34	29
45	27
42	25
32	24
37	14.5
46	41.7
43	32.5
45	47
80	159
39	32
40	28

6. In a study of pro-social message recall from the television, it was hypothesized that the more messages seen (measured with hours of TV viewing per day), the more messages the respondent would recall. Does this data support this hypothesis?

Scores		Scores (cont)	
Hrs TV/Day	Message Recall	Hrs TV/Day	Message Recall
7	3	4	4
6	2	5	1
3	2	2	2
2	0	6	5
6	5	7	3
5	4	9	4
4	3	4	2
6	2	4	1
7	3	7	5
3	2		

7. Does religiosity decrease with income? Are people with higher incomes likely to go to fewer services? Income measured in 1/1000 of value.

Income	No. of Relig services/Year
44	12
35	18
32	24
29	24
34	12
50	6
62	1
11	52
29	6
27	36
25	2
24	1
32	0
41	4
32	12
47	2
159	1
32	12
18	26

8. A professor noticed he had many married students in his class, some with children. Being a father himself, he wondered what effect the children would have on study habits of his married students. What do the data tell us about it?

No. of Children	Hrs of Study/Wk
4	16
0	10
2	12
2	11
1	14
6	20
0	10
1	12
2	15
4	16
3	14
2	17
0	9
0	6
1	10

9. Do people who live near the city take advantage of all the things happening there more so than people who live further away?

Proximity to City	No.Cultural Events/yr
5	25
12	18
25	12
40	2
17	3
10	24
19	6
28	2
36	0
6	3
17	4
28	2
19	1
10	7
27	1

10. It seemed to a researcher that as children grow older, there were fewer and fewer family activities. Does our data support this hypothesis?

Age of Child.	No. Family Activities/Yr
4	4
11	2
15	0
3	12
10	1
13	0
8	36
6	18
5	27
17	0
16	1
14	4
10	23
7	20

11. It has been said that if you want to get a lot accomplished, stay very busy. Do people who are very busy get more done per day than people who are not?

Hrs of Free time/Day	No. of Activities/Wk
5	10
12	2
25	0
40	0
17	2
11	5
19	4
28	2
36	3
6	5
17	2
28	2
19	0
10	7
27	2
11	5
15	3
3	12
10	16
13	7
8	8
6	9
5	10

12. Is television associated with violent acts in children? The hours of TV watching per day and the number of violent acts per day for 16 children are below.

Scores		Scores (cont)	
No. of TV Hrs/Day	No. of Violent Acts	No. of TV Hrs/Day	No. of Violent Acts
2	0	5	4
3	1	1	1
4	3	3	0
2	0	7	0
6	4	8	4
7	3	2	2
5	7	4	5
4	2	7	4

DETAILED ANSWERS

1.
$\sum X =$ 97
$\sum X^2 =$ 837
$\sum Y =$ 1238
$\sum Y^2 =$ 98196
$\sum XY =$ 7968

$\bar{X}=6.0625$ $\bar{Y}= 77.375$

a = 66.1
b = 1.85
r = 0.60
$r^2 =$ 0.36

There is a strong, positive correlation between hours of study and exam grade; 36% of the variance in grades is explained by hours studied. For every one hour that a person studies, we predict an increase in his/her grade of approximately 1.85 points.

2.
$\sum X =$ 306
$\sum Y =$ 805.6
$\sum XY =$ 13703
$\sum X^2 =$ 4990
$\sum Y^2 =$ 56643.4

$\bar{X}=15.3$ $\bar{Y}= 40.28$ ($40,280)

a = -28.1
b = 4.5
r = 0.50
$r^2 =$ 0.25

There is a moderately strong positive association between income and education. 25% of the variance in income can be explained by education. For each additional year of education, a person can expect an increase in salary of $4,500.

3.
$\sum X =$ 782
$\sum X^2 =$ 66752
$\sum Y =$ 50
$\sum Y^2 =$ 200
$\sum XY =$ 1735

$\bar{X}$ =48.9 $\bar{Y}$ = 3.1

a = 4.3
b = -0.02
r = -0.6
r^2 = 0.36

There is a negative, moderately strong association between income and number of children. As income increases, the number of children a couple plans to have decreases; 36% of the variance in number of children is explained by income. For every one unit change in income, we can expect the number of children to decrease by 0.02.

4.
$\sum X$ = 196
$\sum X^2$ = 2830
$\sum Y$ = 298
$\sum Y^2$ = 6498
$\sum XY$ = 4236

$\bar{X}$=14

$\bar{Y}$=21.3

a = 10.9
b = 0.7
r = 0.6
r^2 = 0.36

There is a moderately strong positive association between education and age wed. For every one year increase in education, we can expect the age of first married to increase by 0.7.

5.
$\sum X$ = 822
$\sum X^2$ = 38634
$\sum Y$ = 831.2
$\sum Y^2$ = 67693.6
$\sum XY$= 45258.2

$\bar{X}$=43.3

$\bar{Y}$=$43,700

a = -87.2
b = 3.02
r = 0.9
r^2 = 0.81

There is a very strong, positive association between hours worked per week and income, with 81% of the variance explained in income. For every additional hour worked, income is expected to increase by $3,020

6.

$\sum X =$ 97
$\sum X^2 =$ 561 $\bar{X} = 5.1$
$\sum Y =$ 53
$\sum Y^2 =$ 185 $\bar{Y} = 2.8$
$\sum XY =$ 299

a = 0.58
b = 0.43
r = 0.57
$r^2 =$ 0.32

There is a moderate positive association between hours of television watched and number of messages seen. For every hour of TV watched, an additional 0.43 messages are seen.

7.

$\sum X =$ 763
$\sum X^2 =$ 47985 $\bar{X} = 40.2$
$\sum Y =$ 251
$\sum Y^2 =$ 6827 $\bar{Y} = 13.2$
$\sum XY =$ 6837
a = 20.7
b = -0.19
r = -0.42
$r^2 =$ 0.18

There is a moderate, negative relationship between income and religious services. As income increases, number of religious services attended decreases. Only 18% of the variance in number of religious services attended is explained by income, though.

8.

$\sum X =$ 28
$\sum X^2 =$ 96 $\bar{X} = 1.9$
$\sum Y =$ 192
$\sum Y^2 =$ 2644 $\bar{Y} = 12.8$
$\sum XY =$ 436

a = 9.5
b = 1.8
r = 0.9
$r^2 =$ 0.81

There is a strong, positive association between number of children and hours of study per week.

As the number of children increases, so too do the hours of study. In fact, for every extra child, the professor can expect the student to study 1.8 hours more per week.

9.

$\sum X =$ 299
$\sum X^2 =$ 7523 $\bar{X}=19.9$
$\sum Y =$ 110
$\sum Y^2 =$ 1802 $\bar{Y}=7.3$
$\sum XY =$ 1440

a = 16.9
b = -0.48
r = -0.60
$r^2 =$ 0.36

Yes, there is a negative association between these two variables which indicates that the closer one lives to the city center, the more events s/he will attend, with 36% of the variance in number of events explained by proximity to city. For each additional mile away from the city center a person lives, we can predict him/her to attend approximately 0.48 fewer events.

10.

$\sum X =$ 139
$\sum X^2 =$ 1655 $\bar{X}=9.9$
$\sum Y =$ 148
$\sum Y^2 =$ 3460 $\bar{Y}=10.6$
$\sum XY =$ 1057

a = 25.5
b = -1.5
r = -0.57
$r^2 =$ 0.32

There is a moderately strong, negative association between age of child and number of family events. The data do support the hypothesis that as children grow older, families spend less time together.

11.

$\sum X=371$
$\sum X^2=8293$ $\bar{X}=16.1$
$\sum Y=116$
$\sum Y^2=976$ $\bar{Y}=5.0$
$\sum XY=1202$

a = 9.7
b = -0.29
r = -0.7

$r^2 =$ 0.49

There is a fairly strong, negative association between hours of free time and number of activities, which indicates that people who are busier do get more done. 49% of the variance in number of activities is explained by hours of free time per day.

12.

$\sum X =$ 70

$\sum X^2 =$ 376 $\bar{X} = 4.4$

$\sum Y =$ 40

$\sum Y^2 =$ 166 $\bar{Y} = 2.5$

$\sum XY =$ 208

a = 0.43
b = 0.47
r = 0.49
$r^2 =$ 0.24

There is a moderately strong, positive association between television viewing and violence among children, with 24% of the variance in violence explained by hours of television watching. For every hour of TV viewing, we predict an increase of 0.47 in violent acts.

CHAPTER SEVENTEEN
ELABORATING BIVARIATE TABLES

Most often in social science research we are not concerned with only two variables, but instead with multiple variables, or what is called **multivariate** problems. As such, the tests we have learned up to now are not sufficient for these types of questions. Therefore, we need to use techniques which take into account more than two variables.

When we introduce a third variable into our analysis, we often call this variable Z (do not confuse this with a Z score or a critical value), which refers to the relationship between three variables: X, Y, and Z. This third variable is also called a control variable, which means we are interested in the effects of X (the independent variable) on Y (the dependent variable), controlling for the effects of this third variable (Z). "Controlling for" is another way of saying holding constant. This is all done by way of bivariate tables and partial tables for nominal and ordinal variables. We compare the lambda or gamma from the bivariate table (the association between the two variables) with the lambdas or gammas of the partial tables which control for the third variable.

Let's say we are interested in the relationship between job satisfaction and marital happiness. We find that there is a positive relationship between these two variables (as job satisfaction increases, so too does marital happiness), but we suspect that this is true for men, but not for women. Thus, we control for, or hold constant, the third variable, gender. Then we look at the relationship between job satisfaction and happiness for men, and the relationship between job satisfaction and happiness for women. What we find is that the relationship between the two variables is positive for men, but does not exist for women. Therefore we conclude that the third variable, gender, has an interaction effect between the first two variables.

When using control variables, there are three types of situations which can occur. First, the relationship between X and Y is the same for all categories of the control variable, and the same as the bivariate relationship, or what is called a **direct** relationship. Second, the relationship between the bivariate relationship of X and Y drops off for all categories of the control variable. This pattern indicates either a **spurious** or an **intervening** relationship. Or, thirdly, the relationship between X and Y <u>changes</u> depending on which category you look at for the control variable; the relationship may increase for one category, but decrease for another category, or it may increase for one and drop off to zero for another. This pattern is called **interaction**.

A few examples might make it easier to remember these different patterns. Direct relationships are the easiest: there is basically no change in the partial tables with the control variables from the original bivariate table. This indicates that the control variable has little or no effect on the relationship between X and Y. For example, what if you found a positive association between education and income, and decided to control for gender. If you found a positive relationship of about the same strength for males and for females as you had found in the bivariate relationship, then you would conclude that gender has little or no effect on the bivariate relationship, and that the relationship between education and income is a direct relationship.

The second possible relationship is one in which the bivariate relationship drops off to about zero when the control variable is taken into account. This relationship could indicate one of two patterns. The first, a **spurious** relationship is when the control variable is antecedent, or comes before, the bivariate relationship, and therefore washes out that relationship. For example, a researcher found that increased ice cream sales were positively associated with increased drowning incidents. Therefore, he concluded that eating ice cream increases the likelihood of drowning. Another researcher came along and decided to control for the time of year, specifically summer and winter. What he found, then, was that when you control for the seasons, the relationship between ice cream sales and drowning drops off. More specifically, summertime causes ice cream sales to increase, and summertime causes incidents of drowning to increase. The two variables, ice cream sales and drowning, then, are not related once you control for time of year.

A second possible relationship for this pattern is when the control variable intervenes (thus called **intervening**) between the independent and dependent variable, again washing out the relationship between the two. For example, a researcher found that age has a positive association with income: the older one is, the more money he/she makes. She then decided to control for education and found that once education was controlled for the relationship between age and income disappeared. She concluded that as age increases, so too does education, which then increases income.

Deciding whether the pattern is spurious or intervening is based on theoretical and logical grounds. Think to yourself: do ice cream sales cause it to be summer which in turn causes more drownings (which would indicate an intervening relationship), or is summer an antecedent variable which causes increased ice cream sales and increased cases of drowning (in which case the relationship would be spurious)? The pattern must follow some sort of logical order.

The third possible relationship between the bivariate tables and partial tables is one in which the different categories of the control variable have different effects on the dependent variable. That is, let's say you found a positive association between television viewing and violence in children, but then decided to control for income. You divided the children into two groups, low income and high income, and then analyzed the relationship between TV viewing and violence for both groups. The pattern of the relationship was such that for the low income children there was a positive association (as you'd seen in the original bivariate table), but for high income children, there was no relationship (a gamma of 0.01). Therefore, you conclude that yes, the control variable does have an effect on this relationship which is called an **interaction** effect.

Another possible pattern which might be seen in the interaction relationship is a positive relationship for one category and a negative relationship for another category. For example, let's say you found a positive relationship between hours per week at a paying job and marital satisfaction (that is, the more hours per week one spent at his/her job, the more satisfied she/he would be in his/her marriage). Since you hypothesized that this might be different for men and women, you controlled for gender and found that in fact it's true: for men, there is a positive relationship between hours at work and marital satisfaction, but for women there is a negative relationship between these two variables. Therefore, gender does affect the relationship between hours worked and job satisfaction, but differently for both sexes. This again is also called interaction.

MULTIPLE-CHOICE QUESTIONS

1. Multivariate indicates we are interested in __________ variables.
   a. one
   b. two
   c. three
   d. three or more

2. For nominal and ordinal multivariate analysis, we compare ______________ with ______________.
   a. bivariate tables, multivariate statistics
   b. bivariate gammas, partial lambdas
   c. bivariate tables, partial tables
   d. correlation coefficient, partial tables

3. In multivariate analysis, ________ indicates the control variable.
   a. X
   b. Y
   c. Z
   d. C

4. Controlling for a relationship means that we are ____________ the control variable.
   a. manipulating
   b. holding constant
   c. withholding
   d. relinquishing

5. If both partial tables are significantly less than the bivariate table, we might describe this relationship as:
   a. direct
   b. spurious
   c. interaction
   d. specification

6. If the test statistics for both partial tables drop off to basically zero, and you suspect the control variable to be antecedent to the independent variable, then you would conclude that the control variable is ____________.
   a. direct
   b. spurious
   c. intervening
   d. interaction

7. If the test statistics for both partial tables drop off to basically zero, and you suspect the control variable to come between the independent and dependent variables, then you would conclude that the relationship is ______________.
   a. direct
   b. spurious
   c. intervening
   d. interaction

8. If there is a mixed reaction to the control variable, you might conclude that the relationship is ____________.
   a. direct
   b. spurious
   c. intervening
   d. interaction

9. If we had a situation in which, when controlling for ethnicity, whites in our sample had a similar pattern as the bivariate relationship, but the relationship for the non-whites dropped off to zero, then we would conclude that the relationship is ____________.
   a. direct
   b. spurious
   c. intervening
   d. interaction

10. Spurious relationships are also called:
    a. interpretation
    b. explanation
    c. specification
    d. replication

11. Direct relationships are also called:
    a. interpretation
    b. explanation
    c. specification
    d. replication

12. Interaction relationships are also called:
    a. interpretation
    b. explanation
    c. specification
    d. replication

13. Intervening relationships are also called:
a. interpretation
b. explanation
c. specification
d. replication

14. Partial gammas are used:
a. in place of partial tables
b. to indicate the strength of the control variable
c. to indicate the strength of the bivariate relationship once the effects of the control variable has been removed
d. to indicate the strength of the bivariate relationship when no control variable can be determined

15. Deciding if a relationship is either spurious or intervening depends on:
a. bivariate table
b. partial tables
c. partial gamma
d. theory and logic

WORK PROBLEMS

1. A researcher found there was a positive association between noise tolerance and tolerance for children (G=0.54). What happens to the relationship if she controls for gender?

MALES

TOLERANCE FOR CHILD.	NOISE TOLERANCE LOW	MODERATE	HIGH	
LOW	13	5	2	20
MODERATE	6	18	3	27
HIGH	2	2	9	13
	21	25	14	60

FEMALES

TOLERANCE FOR CHILD.	NOISE TOLERANCE LOW	MODERATE	HIGH	
LOW	8	7	3	18
MODERATE	10	7	7	24
HIGH	8	10	13	31
	26	24	23	73

2. When a researcher analyzed the relationship between media consumption (hours of TV watched per day) and knowledge of family planning methods (which were described in pro-social spots on TV), he found a strong, positive association (G=0.79). How does the relationship change if we control for education level?

HIGH EDUCATION

KNOWL. OF FAMILY PLANNING	MEDIA CONSUMPTION LOW	HIGH	
LOW	74	68	142
MODERATE	85	72	157
HIGH	81	77	158
	240	217	457

LOW EDUCATION

KNOWL. OF FAMILY PLANNING	MEDIA CONSUMPTION LOW	HIGH	
LOW	78	85	163
MODERATE	68	83	151
HIGH	72	79	151
	218	247	465

3. Compute a partial gamma (G_p) for Question two, above. What does this tell us about the relationship between X and Y?

4. We found earlier (chapter 15) that there was a positive association between the popularity of an older sibling and the popularity of a younger sibling (G=0.44). Face it: you are only as hip as your surroundings--so, how does this relationship change if we control for the age of parents?

YOUNG PARENTS

POPULARITY OF YOUNGER SIBLING	POPULARITY OF OLDER SIBLING LOW	HIGH	
LOW	9	7	16
HIGH	6	10	16
	15	17	32

OLD PARENTS

POPULARITY OF YOUNGER SIBLING	POPULARITY OF OLDER SIBLING LOW	HIGH	
LOW	8	6	14
HIGH	6	10	16
	14	16	30

5. Compute a partial gamma (G_p) for Question four, above. What does this tell us about the relationship between X and Y?

6. A professor found a positive association (0.40) between effort put into schoolwork and grades. She wondered if IQ might have an effect on this relationship. Analyze the data below for the effect, if any.

HIGH IQ

GRADES	EFFORT IN SCHOOLWORK: LOW	HIGH	
LOW	7	7	14
HIGH	9	13	22
	16	20	36

LOW IQ

GRADES	EFFORT IN SCHOOLWORK: LOW	HIGH	
LOW	16	12	28
HIGH	10	16	26
	26	28	54

7. Marital patterns were analyzed earlier in terms of SES of the respondent's birth family and the SES of his/her spouse's family; a positive association was found (0.60). A researcher hypothesizes that women more often than not marry up and men marry down (the social ladder). Does the data below support an argument for controlling for gender?

MALES

SES OF SPOUSE'S FAMILY	SES OF FAMILY: UPPER	MIDDLE	LOWER	
UPPER	45	100	10	155
MIDDLE	47	450	21	518
LOWER	33	450	94	577
	125	1000	125	1250

FEMALES

SES OF SPOUSE'S FAMILY	SES OF FAMILY: UPPER	MIDDLE	LOWER	
UPPER	112	330	24	466
MIDDLE	11	560	63	634
LOWER	2	110	38	150
	125	1000	125	1250

8. Compute a partial gamma (G_p) for Question seven, above. What does this tell us about the relationship between X and Y?

DETAILED ANSWERS

1. Males:
N_s=704
N_d=144
G=0.66

Females:
N_s=757
N_d=437
G=0.27

The bivariate gamma was 0.54. The partial table for males had an increase in gamma (0.66), while for females there was a decrease (0.27). Therefore there is probably an interaction effect happening between tolerance of noise and tolerance for children with gender. There appears to be a stronger association between these two variables for males than for females.

2. High Education:
N_s=17571
N_d=17120
G=0.01

Low Education:
N_s=18008
N_d=17876
G=0.004

Since the gammas in both partial tables dropped off to basically zero, we would conclude that the control variable does have an effect on the bivariate relationship, either spurious or intervening. Because media consumption is probably influenced by education and knowledge of family planning is probably influenced by education, we could conclude that the control variable is antecedent, therefore, the relationship is spurious.

3.
$\sum N_s$ = 17571 + 18008 = 35579
$\sum N_d$ = 17120 + 17876 = 34995

$$G_p = \frac{\sum N_s - \sum N_d}{\sum N_s - \sum N_d} = \frac{35579 - 34995}{35579 - 34995} = 0.008$$

The partial gamma indicates that once education is controlled for, the relationship between X and Y (media consumption and knowledge of family planning methods) doesn't exist.

4. Young Parents:
N_s=90
N_d=42
G=0.36

Old Parents:
N_s=80
N_d=36
G=0.40

The gamma for the bivariate table is essentially the same as the gammas for both partial tables. Therefore, the age of parents has no effect on the relationship between older sibling popularity and younger sibling popularity.

5.
$\sum N_s=90+80=170$
$\sum N_d=42+36=78$

$$G_p = \frac{\sum N_s - \sum N_d}{\sum N_s - \sum N_d} = \frac{170 - 78}{170 + 78} = 0.37$$

The partial gamma of 0.37 indicates that there is virtually no change in the relationship between popularity of older and younger siblings even after the effects of age of parents is controlled for.

6. High IQ
N_s=91
N_d=62
G=0.18

Low IQ
N_s=256
N_d=120
G=0.36

The original association between the bivariate relationship was 0.40, and there appears to be a quite a change for those students with high IQs, but not for those students with low IQs. This data tells us that effort has a positive, moderate association with grades for low IQ students, but a very low, albeit positive, association for students with high IQs. Since the relationship drops off quite a bit for one group, but not the other, we could conclude that the variable IQ has an interaction effect on the bivariate relationship.

7. Males
N_s=125043
N_d=42793

G=0.49

Females
N_s=142590
N_d=28858
G=0.66

The relationship between X and Y is virtually unchanged in the two partial tables, therefore I would conclude that there is a direct relationship between SES of birth family and the SES of spouses family. Gender appears to have very little effect on this relationship.

8.
$\sum N_s$=267633
$\sum N_d$=71651

$$G_p = \frac{\sum N_s - \sum N_d}{\sum N_s - \sum N_d} = \frac{267633 - 71651}{267633 + 71651} = 0.58$$

The partial gamma indicates that the relationship between X and Y is unchanged when we control for gender, therefore there is a direct relationship between X and Y.

CHAPTER EIGHTEEN
PARTIAL CORRELATION AND MULTIPLE REGRESSION AND CORRELATION

Now that you have been introduced to the wonders of multivariate techniques, the fun begins! In this chapter multivariate techniques for interval-ratio variables are introduced. Specifically, partial correlation, multiple regression, and multiple correlation are discussed.

Partial correlation uses the same basic logic as partial tables discussed in chapter 17. The relationship between X and Y is analyzed, then the effects of the third variable, Z, are controlled for. Because interval-ratio data is continuous, we do not use bivariate tables and the like, but, instead, correlations. When we calculated the correlation between two variables, we were computing Pearson's r. These correlations are also called **zero-order correlations**, and refer to the correlation between two variables. In order to control for the effects of the third variable, Z, we need to use **partial correlations**. Partial correlations will help us examine the relationship between X and Y controlling for Z (symbolically given as $r_{yx.z}$).

The zero-order correlation (r_{yx}) will then be compared to the partial correlation ($r_{yx.z}$), in the same manner as we compared the bivariate gamma to the gammas of each of the partial tables. Likewise, there are three patterns we are concerned with. First, if the zero-order correlation is approximately equal to the partial correlation, then we conclude that the control variable has little or no effect on the bivariate relationship, therefore, there is a direct relationship between X and Y.

The second possible pattern is if the zero-order correlation is much larger than the partial correlation ($r_{yx} > r_{yx.z}$), then we conclude that the pattern is either spurious or intervening, meaning that the relationship between X and Y drops off once the third variable is controlled for. Again, as was the case in nominal and ordinal data, the conclusion that the relationship is spurious or that it is intervening must be based on theory or logical ordering of the variables.

The third possible pattern is if the zero-order correlation is much smaller than the partial correlation ($r_{yx} < r_{yx.z}$). If this is the case then we conclude that both X and Z are independent variables, both causing Y, and therefore we will want to find out which variable has the largest effect on Y. Thus, we go to the next step: multiple regression.

There are two main parts to multiple regression. First, as we saw in least-squares regression (chapter 16), we can make predictions of the value for the dependent variable with the regression equation:

$$Y = a + b_1 X_1 + b_2 X_2$$

In this equation, we need to find a (the Y-intercept), and b_1 and b_2 (the partial slopes for each of the independent variables). We can then plug the values of X1 and X2 into the equation, as we did in chapter 16, and make a prediction for the value of Y.

The second part to multiple regression is assessing the effects of the two independent variables: which variable X (now called X_1) or Z (now called X_2) has the largest effect on Y? In order to determine this, we need to convert all variables into Z scores.

Think way back to chapter five. Remember how we can convert items into Z scores in order to change the scales? Think back to the example of temperatures measured at Celsius and Fahrenheit: if we convert the temperature outside from our usual scale of Fahrenheit to the scale of Celsius, it is still the same temperature, now with a freezing point of zero instead of 32°.

This is the same logic we are using when we want to assess the effects of X_1 and X_2 on Y. Since the variables are usually measuring something different (such as hours worked per week, education, and income), it is difficult to compare them (the old apples and oranges problem). Instead, we convert all the data to Z scores, so each distribution has a mean of zero and a standard deviation of one, and then compare themThe simplest way to do this, then is to use the following equation:

$$Z_y = a_z + b^*_1 Z_1 + b^*_2 Z_1$$

in which b^*_1 and b^*_2 are the standardized slopes for each of the two independent variables, or what are called **beta weights**. With these beta weights, you can assess the effect of each independent variable on the dependent variable separately to determine which has the strongest effect. Beta weights can range from -1 to +1, so if $b^*_1=-0.76$ and $b^*_2=+0.66$, then X_1 has the strongest effect (-0.76 > +0.66).

Finally, PHEW!, the effects of ALL the independent variables together can be assessed by using **multiple correlation** (**R**) and the **coefficient of multiple determination** (**R^2**). When we were analyzing the association between two interval-ratio variables, we used Pearson's r and r^2, or the coefficient of determination. The coefficient of determination (r^2) was a PRE-type measure which told us the amount of variation explained in the dependent variable by the independent variable. Similarly, the coefficient of multiple determination (**R^2**) is the amount of variation explained in the dependent variable by the first independent variable (X_1) and by the second independent variable (X_2) after the shared variation has been removed. More succinctly, R^2 is the amount of variation explained in the dependent variable by all of the independent variables.

Thus, if we have an R^2 of 0.25, then we conclude that 25% of the variation (or variance) in the dependent variable is explained by the independent variables.

Sometimes you might see correlations in a matrix form, called a correlation matrix, with each variable label across the top and down the side, and the correlation between each variable in the corresponding cell. Along the diagonal (in cell X_1, X_1 for example) is the correlation of each variable with itself, which, logically is 1.0. Many times these matrices will be only half complete since the matrix is symmetric (the correlation for X_1 and X_2 is the same as the correlation for X_2 and X_1). A three variable correlation matrix may look something like this:

	X_1	X_2	Y
X_1	1.0	-0.31	0.56
X_2		1.0	0.76
Y			1.0

Thus, the correlation between X_1 and X_2 is -0.31, between X_1 and Y is 0.56, and between X_2 and Y is 0.76. Note that the lower triangle of the matrix is incomplete. You can complete it if you'd like with the same numbers that are in the upper triangle, but it would be redundant, so it's not necessary.

The primary reason these matrices are used is for ease of reading the statistics. You need to look at this table only briefly to see the relationship between the three variables. Correlation matrices are not limited to any number of variables, but can get cumbersome when too many variables are included.

MULTIPLE-CHOICE QUESTIONS

1. If we want to control for the effects of a third interval-ratio variable, we use ________________.
   a. partial tables
   b. partial correlations
   c. multiple regression
   d. beta weights

2. Multiple regression is used when you have ________ dependent variable(s) and ________ independent variable(s).
   a. two, two
   b. one, two
   c. one, three
   d. one, more than two

3. If we want to make a prediction of the value of Y with more than one independent variable, we use ________________.
   a. partial tables
   b. partial correlations
   c. multiple regression
   d. beta weights

4. Beta weights are used to ________________.
   a. make predictions
   b. determine the amount of variance explained in the dependent variable by the independent variables
   c. determine the effects of each of the independent variables on the dependent variable
   d. determine causality

5. Partial slopes are ___________________.
a. the slope of the relationship between one independent variable and the dependent variable, controlling for the effects of all other independent variables
b. the amount of variance explained in the dependent variable by the independent variables
c. the correlation for each partial table
d. the combined effects of all the independent variables on the dependent variable

6. The coefficient of multiple determination is ___________.
a. r^2
b. b^*
c. R^2
d. Z_y

7. The coefficient of determination is the ____________________________________.
a. slope of the relationship between a particular independent variable and the dependent variable
b. amount of variation explained in the dependent variable by all independent variables combined
c. combined effects of all the independent variables on the dependent variable
d. slope of the relationship between one independent variable and the dependent variable, controlling for the effects of all other independent variables

8. The symbol for the relationship between X and Y controlling for Z is:
a. R^2
b. b_1^*
c. $r_{yx.z}$
d. $r_{zx.y}$

9. If we have two partial slopes, the first -0.26, the second 0.19, we can conclude:
a. the second variable has a stronger effect
b. there is no association between these two variables and the dependent variable
c. the first variable has a stronger effect
d. there is a mixed correlation between the variables

10. The multiple correlation coefficient is used to:
a. determine the slope of the relationship between a particular independent variable and the dependent variable
b. determine the strength of the association between a dependent variable and two or more independent variables
c. determine the amount of variance explained in the dependent variable by the independent variables
d. determine the effects of each of the independent variables on the dependent variable

11. In the standardized least-squares equation, ______________ are used to normalize the distributions.
a. partial slopes
b. partial correlations
c. Z scores
d. all of the above

12. Zero-order correlations are used to indicate the correlation between ____________.
a. two variables
b. ordinal variables
c. one ordinal and one interval-ratio variable
d. beta weights

13. Unstandardized partial slopes indicate:
a. the effect of a particular independent variable on the dependent variable
b. the amount of variance explained in the dependent variable
c. the predicted value of Z
d. the amount change in Y for a unit change in X

14. Z_1 is:
a. the Z score for the first dependent variable
b. the critical value for the amount of variance explained in the dependent variable
c. the standardized score for the first independent variable
d. the standardized score for the second independent variable

15. The term Z_y is used to denote:
a. the critical value for the amount of variance explained in the dependent variable
b. the standardized regression equation
c. the amount of variance explained in the dependent variable
d. the beta weights

WORK PROBLEMS

1. A researcher was interested in the relationship between education and income, and although they had a fairly strong, positive relationship, she thought that job prestige might explain more of the variance. Below are the correlations, means, and standard deviations for the data.

	EDUC (X_1)	PRESTIGE (X_2)	INCOME (Y)
EDUC (X_1)	1.0	0.46	0.54
PRESTIGE (X_2)		1.0	0.32
INCOME (Y)			1.0

$\bar{X}_1=13.5$ $\bar{X}_2=42.1$ $\bar{Y}=\$32,590$

$s_1= 3.7$ $s_2=15.6$ $s_y=\$8,970$

a. Predict the income for someone who has 15 years of education and a prestige score of 63.

b. Which independent variable has the strongest effect on the dependent variable?

c. How much variance is explained in the dependent variable by the independent variable?

2. Quality of life of AIDS patients was hypothesized to be affected by health status and the length of time a person was HIV positive. The data are presented below.

	HEALTH (X_1)	TIME HIV+ (X_2)	QUALITY OF LIFE (Y)
HEALTH (X_1)	1.0	-0.31	0.63
TIME HIV+ (X_2)		1.0	-0.43
QUALITY OF LIFE (Y)			1.0

$\bar{X}_1$=10.8 $\bar{X}_2$=53.3 $\bar{Y}$=42.3
s_1= 6.1 s_2=34.1 s_y=29.6

a. Which independent variable has the strongest effect on the dependent variable?

b. How much variance is explained in the dependent variable by the independent variable?

3. In a study of knowledge of family planning, a researcher found that two variables were correlated with knowledge: hours of television watched per day and cosmopolitanism (worldliness).

	TV/day (X_1)	COSMO (X_2)	KNOWL. FAM PLAN (Y)
TV/day (X_1)	1.0	0.05	-0.47
COSMO. (X_2)		1.0	0.56
KNOWL. FAM PLAN (Y)			1.0

$\bar{X}_1$= 4.3 $\bar{X}_2$=27.6 $\bar{Y}$=8.3
s_1= 1.5 s_2=18.3 s_y=2.9

a. Discuss each of the relationships in the correlation matrix.

b. Which variable has the strongest effect on the dependent variable?

c. How much variance in knowledge of family planning is explained by these variables?

4. The Wesaloosa dorm guys are still trying to figure out why they aren't getting any dates. They decided to find out what might be the primary reason. Given the data below, what would you conclude?

	HOURS OF STUDY (X_1)	CLEAN-LINESS (X_2)	NO. OF DATES (Y)
HOURS OF STUDY (X_1)	1.0	0.09	-0.59
CLEAN-LINESS (X_2)		1.0	-0.37
NO. OF DATES (Y)			1.0

$\bar{X}_1 = 3.5$ $\bar{X}_2 = 56$ $\bar{Y} = 2.1$
$s_1 = 1.3$ $s_2 = 23$ $s_y = 1.6$

5. A criminologist was interested in the relationship between truancy and criminal acts committed by teenagers. He hypothesized that socioeconomic status might also have an effect.

a. Which variable has the strongest effect?

b. How much variance is explained in criminal activity by these two variables?

	TRUANCY (X_1)	SES (X_2)	CRIMINAL ACTS (Y)
TRUANCY (X_1)	1.0	-0.35	0.68
SES (X_2)		1.0	-0.22
CRIMINAL ACTS (Y)			1.0

$\bar{X}_1 = 16$ $\bar{X}_2 = 46.2$ $\bar{Y} = 3.1$
$s_1 = 23$ $s_2 = 24.3$ $s_y = 2.3$

DETAILED ANSWERS

1. a. We predict that a person with 15 years of education and job with a prestige score of 63 will make approximately $35,537 per year.

   b. $b^*_1=0.50$
   $b^*_2=0.09$

   Thus, the first independent variable, education, has the strongest effect on the dependent variable, income.

   c. $R^2=0.30$ -- 30% of the variance in income is explained by education and occupational prestige, with 29% of that variance explained by education alone.

2. a. $b^*_1=0.55$
   $b^*_2=-0.26$

   Health status appears to have the stronger effect on quality of life. Note that health is positive, indicating that healthier individuals have higher quality of life, while length of time is negative, indicating that the longer one is HIV positive, the lower his/her quality of life.

3. a. Hours of TV watching per day has a negative relationship with knowledge of family planning, indicating that as TV viewing increases, knowledge of family planning decreases. Cosmopolitanism and knowledge of family planning have a moderately strong, positive association.

   b. $b^*_1=-0.50$
   $b^*_2=0.58$

   Cosmopolitanism appears to have the strongest effect on knowledge of family planning.

   c. The amount of variation explained in the dependent variable by all of the independent variables is 56%, 22% of which is explained by the first independent variable, TV viewing, alone.

4. There is a moderately strong, negative association between hours of study per week and number of dates, and a weak-moderate negative correlation between number of dates per week and cleanliness. Hours spent studying per week seems to have the strongest effect on number of dates ($b^*_1=-0.56$), with 45% of the variance in number of dates explained by these two variables. Unfortunately, that leaves an additional 55% of the variance <u>un</u>explained. Keep trying, guys!

5. Truancy has the strongest effect on the criminal activity variable for teenagers (b^*_1=0.57), with 34% of the variance explained in the dependent variable by both of the independent variables.

MULTIPLE-CHOICE ANSWERS

CHAPTER 1

1.	b	9.	b
2.	d	10.	c
3.	c	11.	d
4.	a	12.	c
5.	c	13.	b
6.	b	14.	c
7.	a	15.	b
8.	c		

CHAPTER 2

1.	d	9.	a
2.	c	10.	a
3.	c	11.	d
4.	d	12.	b
5.	d	13.	c
6.	a	14.	d
7.	b	15.	a
8.	a		

CHAPTER 3

1.	d	9.	a
2.	d	10.	d
3.	b	11.	d
4.	a	12.	c

5.	c	13.	b
6.	d	14.	a
7.	b	15.	c
8.	a		

CHAPTER 4

1.	c	9.	a
2.	a	10.	c
3.	c	11.	d
4.	a	12.	a
5.	c	13.	d
6.	b	14.	d
7.	a	15.	d
8.	d		

CHAPTER 5

1.	c	9.	b
2.	d	10.	c
3.	b	11.	d
4.	d	12.	d
5.	b	13.	c
6.	a	14.	a
7.	b	15.	c
8.	c		

CHAPTER 6

1.	d	9.	c
2.	d	10.	c
3.	c	11.	a
4.	a	12.	d
5.	c	13.	b
6.	a	14.	a
7.	d	15.	d
8.	b		

CHAPTER 7

1.	d	9.	a
2.	d	10.	c
3.	a	11.	d
4.	d	12.	a
5.	d	13.	b
6.	b	14.	b
7.	c	15.	c
8.	d		

CHAPTER 8

1.	b	9.	c
2.	c	10.	c
3.	d	11.	d
4.	a	12.	a

5.	c	13.	b
6.	c	14.	b
7.	a	15.	a
8.	b		

CHAPTER 9

1.	d	9.	d
2.	b	10.	a
3.	d	11.	b
4.	d	12.	b
5.	c	13.	a
6.	b	14.	d
7.	d	15.	c
8.	d		

CHAPTER 10

1.	b	9.	b
2.	c	10.	c
3.	d	11.	d
4.	a	12.	b
5.	d	13.	d
6.	b	14.	a
7.	c	15.	d
8.	c		

CHAPTER 11

1.	d	6.	c
2.	c	7.	a
3.	d	8.	b
4.	b	9.	d
5.	a	10.	a

CHAPTER 12

1.	b	9.	c
2.	c	10.	d
3.	d	11.	b
4.	a	12.	d
5.	d	13.	d
6.	b	14.	c
7.	a	15.	a
8.	a		

CHAPTER 13

1.	b	9.	d
2.	a	10.	b
3.	b	11.	a
4.	a	12.	d
5.	a	13.	c
6.	b	14.	b
7.	b	15.	c

8.	a		

CHAPTER 14

1.	a	9.	d
2.	c	10.	c
3.	b	11.	a
4.	c	12.	b
5.	d	13.	b
6.	c	14.	d
7.	d	15.	d
8.	a		

CHAPTER 15

1.	b	9.	c
2.	d	10.	b
3.	a	11.	c
4.	b	12.	c
5.	c	13.	c
6.	d	14.	a
7.	b	15.	b
8.	a		

CHAPTER 16

1.	c	9.	c
2.	d	10.	a
3.	b	11.	d

4.	a	12.	b
5.	d	13.	b
6.	c	14.	b
7.	a	15.	b
8.	c		

CHAPTER 17

1.	d	9.	d
2.	c	10.	b
3.	c	11.	d
4.	b	12.	c
5.	b	13.	a
6.	c	14.	c
7.	c	15.	d
8.	d		

CHAPTER 18

1.	b	9.	c
2.	d	10.	b
3.	c	11.	c
4.	c	12.	a
5.	a	13.	d
6.	d	14.	c
7.	c	15.	b
8.	c		